Classics in Mathematics

Shoshichi Kobayashi   Transformation Groups in Differential Geometry

Shoshichi Kobayashi

Transformation Groups in Differential Geometry

Reprint of the 1972 Edition

Springer

Shoshichi Kobayashi
Department of Mathematics, University of California
Berkeley, CA 94720-3840
USA

Originally published as Vol. 70 of the
Ergebnisse der Mathematik und ihrer Grenzgebiete, 2nd sequence

Mathematics Subject Classification (1991):
Primary 53C20, 53C10, 53C55, 32M05, 32J25, 57S15
Secondary 53C15, 53A10, 53A20, 53A30, 32H20, 58D05

ISBN 3-540-58659-8 Springer-Verlag Berlin Heidelberg New York

Photograph by kind permission of George Bergman

CIP data applied for

Printed in Germany

SPIN 11885733 41/3111 - 5 4 3 2 1 – Printed on acid-free paper

Shoshichi Kobayashi

Transformation Groups in Differential Geometry

Springer-Verlag Berlin Heidelberg New York
1972

Shoshichi Kobayashi
University of California, Berkeley, California

AMS Subject Classifications (1970):

Primary 53 C 20, 53 C 10, 53 C 55, 32 M 05, 32 J 25, 57 E 15

Secondary 53 C 15, 53 A 10, 53 A 20, 53 A 30, 32 H 20, 58 D 05

ISBN 3-540-05848-6 Springer-Verlag Berlin Heidelberg New York
ISBN 0-387-05848-6 Springer-Verlag New York Heidelberg Berlin

 Library of Congress Catalog Card Number 72-80361. Printed in Germany. Printing and binding: Universitätsdruckerei H. Stürtz AG, Würzburg

Preface

Given a mathematical structure, one of the basic associated mathematical objects is its automorphism group. The object of this book is to give a biased account of automorphism groups of differential geometric structures. All geometric structures are not created equal; some are creations of gods while others are products of lesser human minds. Amongst the former, Riemannian and complex structures stand out for their beauty and wealth. A major portion of this book is therefore devoted to these two structures.

Chapter I describes a general theory of automorphisms of geometric structures with emphasis on the question of when the automorphism group can be given a Lie group structure. Basic theorems in this regard are presented in §§ 3, 4 and 5. The concept of G-structure or that of pseudo-group structure enables us to treat most of the interesting geometric structures in a unified manner. In § 8, we sketch the relationship between the two concepts. Chapter I is so arranged that the reader who is primarily interested in Riemannian, complex, conformal and projective structures can skip §§ 5, 6, 7 and 8. This chapter is partly based on lectures I gave in Tokyo and Berkeley in 1965.

Contents of Chapters II and III should be fairly clear from the section headings. It should be pointed out that the results in §§ 3 and 4 of Chapter II will not be used elsewhere in this book and those of §§ 5 and 6 of Chapter II will be needed only in §§ 10 and 12 of Chapter III. I lectured on Chapter II in Berkeley in 1968; Chapter II is a faithful version of the actual lectures.

Chapter IV is concerned with automorphisms of affine, projective and conformal connections. We treat both the projective and the conformal cases in a unified manner.

Throughout the book, we use *Foundations of Differential Geometry* as our standard reference. Some of the referential results which cannot be found there are assembled in Appendices for the convenience of the reader.

As its title indicates, this book is concerned with the differential geometric aspect rather than the differential topological or homological aspect of the theory of transformation groups. We have confined ourselves to presenting only basic results, avoiding difficult theorems. To compensate for the omission of many interesting but difficult results, we have supplied the reader with an extensive list of references.

We have not touched upon homogeneous spaces, partly because they form an independent discipline of their own. While we are interested in automorphisms of given geometric structures, the differential geometry of homogeneous spaces is primarily concerned with geometric objects which are invariant under given transitive transformation groups. For the convenience of the reader, the Bibliography includes papers on the geometry of homogeneous spaces which are related to the topics discussed here.

In concluding this preface, I would like to express my appreciation to a number of mathematicians: Professors Yano and Lichnerowicz, who interested me in this subject through their lectures, books and papers; Professor Ehresmann, who taught me jets, prolongations and infinite pseudo-groups; K. Nomizu, T. Nagano and T. Ochiai, my friends and collaborators in many papers; Professor Matsushima, whose recent monograph on holomorphic vector fields influenced greatly the presentation of Chapter III; Professor Howard, who kindly made his manuscript on holomorphic vector fields available to me. I would like to thank Professor Remmert and Dr. Peters for inviting me to write this book and for their patience.

I am grateful also to the National Science Foundation for its unfailing support given to me during the preparation of this book.

January, 1972 S. Kobayashi

Contents

I. Automorphisms of G-Structures

1. G-Structures

Let M be a differentiable manifold of dimension n and $L(M)$ the bundle of linear frames over M. Then $L(M)$ is a principal fibre bundle over M with group $\mathrm{GL}(n; \mathbf{R})$. Let G be a Lie subgroup of $\mathrm{GL}(n; \mathbf{R})$. By a *$G$-structure* on M we shall mean a differentiable subbundle P of $L(M)$ with structure group G.

There are very few general theorems on G-structures. But we can ask a number of interesting questions on G-structures, and they are often very difficult even for some specific G. It is therefore essential for the study of G-structures to have familiarity with a number of examples.

In general, when M and G are given, there may or may not exist a G-structure on M. If G is a closed subgroup of $\mathrm{GL}(n; \mathbf{R})$, the existence problem becomes the problem of finding cross sections in a certain bundle. Since $\mathrm{GL}(n; \mathbf{R})$ acts on $L(M)$ on the right, a subgroup G also acts on $L(M)$. If G is a closed subgroup of $\mathrm{GL}(n; \mathbf{R})$, then the quotient space $L(M)/G$ is the bundle with fibre $\mathrm{GL}(n; \mathbf{R})/G$ associated with the principal bundle $L(M)$. It is then classical that the G-structures on M are in a natural one-to-one correspondence with the cross sections

$$M \to L(M)/G$$

(see, for example, Kobayashi-Nomizu [1, vol. 1; pp. 57–58]). The so-called obstruction theory gives necessary algebraic-topological conditions on M for the existence of a G-structure (see, for example, Steenrod [1]).

A G-structure P on M is said to be *integrable* if every point of M has a coordinate neighborhood U with local coordinate system $x^1, \ldots, x^n$ such that the cross section $(\partial/\partial x^1, \ldots, \partial/\partial x^n)$ of $L(M)$ over U is a cross section of P over U. We shall call such a local coordinate system $x^1, \ldots, x^n$ *admissible* with respect to the given G-structure P. If $x^1, \ldots, x^n$ and $y^1, \ldots, y^n$ are two admissible local coordinate system in open sets U and V respectively, then the Jacobian matrix $(\partial y^i/\partial x^j)_{i,j=1,\ldots,n}$ is in G at each point of $U \cap V$.

Proposition 1.1. *Let* K *be a tensor over the vector space* $\mathbf{R}^n$ *(i.e., an element of the tensor algebra over* $\mathbf{R}^n$*) and G the group of linear transformations of* $\mathbf{R}^n$ *leaving* K *invariant. Let P be a G-structure on M and K the tensor field on M defined by* K *and P in a natural manner (see the proof below). Then P is integrable if and only if each point of M has a coordinate neighborhood with local coordinate system* $x^1, \ldots, x^n$ *with respect to which the components of K are constant functions on U.*

Proof. We give the definition of K although it is more or less obvious. At each point x of M, we choose a frame u belonging to P. Since u is a linear isomorphism of $\mathbf{R}^n$ onto the tangent space $T_x(M)$, it induces an isomorphism of the tensor algebra over $\mathbf{R}^n$ onto the tensor algebra over $T_x(M)$. Then K_x is the image of K under this isomorphism. The invariance of K by G implies that K_x is defined independent of the choice of u.

Assume that P is integrable and let $x^1, \ldots, x^n$ be an admissible local coordinate system. From the construction above, it is clear that the components of K with respect to $x^1, \ldots, x^n$ coincide with the components of K with respect to the natural basis in $\mathbf{R}^n$ and, hence, are constant functions.

Conversely, let $x^1, \ldots, x^n$ be a local coordinate system with respect to which K has constant components. In general, this coordinate system is not admissible. Consider the frame $(\partial/\partial x^1, \ldots, \partial/\partial x^n)$ at the origin of this coordinate system. By a linear change of this coordinate system, we obtain a new coordinate system $y^1, \ldots, y^n$ such that the frame $(\partial/\partial y^1, \ldots, \partial/\partial y^n)$ at the origin belongs to P. Then K has constant components with respect to $y^1, \ldots, y^n$. These constant components coincide with the components of K with respect to the natural basis of $\mathbf{R}^n$ since $(\partial/\partial y^1, \ldots, \partial/\partial y^n)$ at the origin belong to P. Let u be a frame at $x \in U$ belonging to P. Since the components of K with respect to u coincide with the components of K with respect to the natural basis of $\mathbf{R}^n$ and, hence, with the components of K with respect to $(\partial/\partial y^1, \ldots, \partial/\partial y^n)$, it follows that the frame $(\partial/\partial y^1, \ldots, \partial/\partial y^n)$ at x coincides with u modulo G and, hence, belongs to P. q.e.d.

Proposition 1.2. *If a G-structure P on M is integrable, then P admits a torsionfree connection.*

Proof. Let U be a coordinate neighborhood with admissible local coordinate system $x^1, \ldots, x^n$. Let ω_U be the connection form on $P|U$ defining a flat affine connection on U such that $\partial/\partial x^1, \ldots, \partial/\partial x^n$ are parallel. We cover M by a locally finite family of such open sets U. Taking a partition of unity $\{f_U\}$ subordinate to $\{U\}$, we define a desired connection form ω by

$$\omega=\sum_U \pi^* f_U \cdot \omega_U,$$

where $\pi\colon P \to M$ is the projection. q.e.d.

In some cases, the converse of Proposition 1.2 is true. For such examples, see the next section.

Let P and P' be G-structures over M and M'. Let f be a diffeomorphism of M onto M' and $f_*\colon L(M) \to L(M')$ the induced isomorphism on the bundles of linear frames. If f_* maps P into P', we call f an *isomorphism* of the G-structure P onto the G-structure P'. If $M=M'$ and $P=P'$, then an isomorphism f is called an *automorphism* of the G-structure P.

A vector field X on M is called an *infinitesimal automorphism* of a G-structure P if it generates a local 1-parameter group of automorphisms of P.

As in Proposition 1.1, we consider those G-structures defined by a tensor K. Then the following proposition is evident.

Proposition 1.3. *Let* K *be a tensor over the vector space* $\mathbf{R}^n$ *and* G *the group of linear transformations of* $\mathbf{R}^n$ *leaving* K *invariant. Let* P *be a* G-*structure on* M *and* K *the tensor field on* M *defined by* K *and* P. *Then*

(1) *A diffeomorphism* $f\colon M \to M$ *is an automorphism of* P *if and only if* f *leaves* K *invariant;*

(2) *A vector field* X *on* M *is an infinitesimal automorphism of* P *if and only if* $L_X K=0$, *where* L_X *denotes the Lie derivation with respect to* X.

We shall now study the local behavior of an infinitesimal automorphism of an integrable G-structure. Without loss of generality, we may assume that $M=\mathbf{R}^n$ with natural coordinate system $x^1, \dots, x^n$ and $P=\mathbf{R}^n \times G$. Let X be a vector field in (a neighborhood of the origin of) $\mathbf{R}^n$ and expand its components in power series:

$$X=\sum \xi^i \frac{\partial}{\partial x^i}$$

$$\xi^i \sim \sum_{k=0}^{\infty} \frac{1}{k!} \sum_{j_1, \dots, j_k=1}^{n} a^i_{j_1 \dots j_k} x^{j_1} \dots x^{j_k},$$

where $a^i_{j_1 \dots j_k}$ are symmetric in the subscripts $j_1, \dots, j_k$. Since X is an infinitesimal automorphism of P if and only if the matrix $(\partial \xi^i/\partial x^j)$ belongs to the Lie algebra $\mathfrak{g}$ of G, we may conclude that X is an infinitesimal automorphism of P if and only if, for each fixed $j_2, \dots, j_k$, the matrix $(a^i_{j_1 j_2 \dots j_k})_{i, j_1=1, \dots, n}$ belongs to the Lie algebra $\mathfrak{g}$. This motivates the following definition.

Let $\mathfrak{g}$ be a Lie subalgebra of $\mathfrak{gl}(n; \mathbf{R})$. For $k=0, 1, 2, \ldots$, let $\mathfrak{g}_k$ be the space of symmetric multilinear mappings

$$t\colon \underbrace{\mathbf{R}^n \times \cdots \times \mathbf{R}^n}_{(k+1)\text{-times}} \to \mathbf{R}^n$$

such that, for each fixed $v_1, \ldots, v_k \in \mathbf{R}^n$, the linear transformation

$$v \in \mathbf{R}^n \to t(v, v_1, \ldots, v_k) \in \mathbf{R}^n$$

belongs to $\mathfrak{g}$. In particular, $\mathfrak{g}_0 = \mathfrak{g}$. We call $\mathfrak{g}_k$ the *k-th prolongation* of $\mathfrak{g}$. The first integer k such that $\mathfrak{g}_k = 0$ is called the *order* of $\mathfrak{g}$. If $\mathfrak{g}_k = 0$, then $\mathfrak{g}_{k+1} = \mathfrak{g}_{k+2} = \cdots = 0$. If $\mathfrak{g}_k \neq 0$ for all k, then $\mathfrak{g}$ is said to be of *infinite type*.

Proposition 1.4. *A Lie algebra $\mathfrak{g} \subset \mathfrak{gl}(n; \mathbf{R})$ is of infinite type if it contains a matrix of rank 1 as an element.*

Proof. Let e be a nonzero element of $\mathbf{R}^n$ and α a nonzero element of the dual space of $\mathbf{R}^n$. Then the linear transformation defined by

$$v \in \mathbf{R}^n \to \langle \alpha, v \rangle e \in \mathbf{R}^n$$

is of rank 1, and conversely, every linear transformation of rank 1 is given as above. Assume that the transformation above belongs to $\mathfrak{g}$. For each positive integer k, we define

$$t(v_0, v_1, \ldots, v_k) = \langle \alpha, v_0 \rangle \langle \alpha, v_1 \rangle \cdots \langle \alpha, v_k \rangle e, \qquad v_i \in \mathbf{R}^n.$$

Then t is a nonzero element of $\mathfrak{g}_k$. q.e.d.

We say that a Lie algebra $\mathfrak{g} \subset \mathfrak{gl}(n; \mathbf{R})$ is *elliptic* if it contains no matrix of rank 1. Proposition 1.4 means that if $\mathfrak{g}$ is of finite order, then it is elliptic.

Each Lie subalgebra $\mathfrak{g}$ of $\mathfrak{gl}(n; \mathbf{R})$ gives rise to a graded Lie algebra $\sum_{k=-1}^{\infty} \mathfrak{g}_k$, where $\mathfrak{g}_{-1} = \mathbf{R}^n$. The bracket of $t \in \mathfrak{g}_p$ and $t' \in \mathfrak{g}_q$ is defined by

$$\begin{aligned}[t, t'](v_0, v_1, \ldots, v_{p+q}) = {} & \frac{1}{p!\,(q+1)!} \sum t(t'(v_{j_0}, \ldots, v_{j_q}), v_{j_{q+1}}, \ldots, v_{j_{p+q}}) \\ & - \frac{1}{(p+1)!\,q!} \sum t'(t(v_{k_0}, \ldots, v_{k_p}), v_{k_{p+1}}, \ldots, v_{k_{p+q}}).\end{aligned}$$

In particular, if $t \in \mathfrak{g}_p$, $p \geqq 0$, and $v \in \mathfrak{g}_{-1} = \mathbf{R}^n$, then

$$[t, v](v_1, \ldots, v_p) = t(v, v_1, \ldots, v_p).$$

We explicitly set $[\mathfrak{g}_{-1}, \mathfrak{g}_{-1}] = 0$. This definition is motivated by the following geometrical consideration. Suppose $t = (a^i_{j_0 \ldots j_p}) \in \mathfrak{g}_p$ and $t' = (b^i_{k_0 \ldots k_q}) \in \mathfrak{g}_q$ in terms of components and consider the corresponding vector fields:

$$X = \frac{1}{(p+1)!} \sum a^i_{j_0 \dots j_p} x^{j_0} \dots x^{j_p} \frac{\partial}{\partial x^i},$$

$$Y = \frac{1}{(q+1)!} \sum b^i_{k_0 \dots k_q} x^{k_0} \dots x^{k_q} \frac{\partial}{\partial x^i}.$$

Then $[X, Y]$ corresponds to $[t, t']$. Thus, the graded Lie algebra $\sum_{k=-1}^{\infty} \mathfrak{g}_k$ may be considered as the Lie algebra of infinitesimal automorphisms $X = \sum \xi^i \frac{\partial}{\partial x^i}$ with polynomial components ξ^i of the flat G-structure $P = \mathbf{R}^n \times G$ on $\mathbf{R}^n$.

For a survey on G-structures, see expository articles of Chern [1], [2]; the latter contains an extensive list of publications on the subject. See also Sternberg's book [1], A. Fujimoto [2], [3], Bernard [1].

The group of automorphisms of a compact elliptic structure or a G-structure of finite type will be shown to be a Lie transformation group (see §§ 4 and 5, respectively). These two cases cover a substantial number of interesting geometric structures whose automorphism groups are Lie groups. By considering G-structures of higher degree, we can bring such structures as projective structures under this general scheme (see § 8 of this chapter and Chapter IV). The group of automorphisms of a bounded domain or a similar complex manifold is also a Lie group (see § 1 of Chapter III), but this does not come under the general scheme. This book does not touch area-measure structures (Brickell [1]), nor pseudo-conformal structures of real hypersurfaces in complex manifolds (Morimoto-Nagano [1], Tanaka [3]) although automorphism groups of these structures are usually Lie groups.

2. Examples of G-Structures

Example 2.1. $G = \mathrm{GL}(n; \mathbf{R})$ and $\mathfrak{g} = \mathfrak{gl}(n; \mathbf{R})$. The Lie algebra $\mathfrak{g}$ contains a matrix of rank 1 and is of infinite type. A G-structure on M is nothing but the bundle $L(M)$ of linear frames and is obviously integrable. Every diffeomorphism of M onto itself is an automorphism of this G-structure and every vector field on M is an infinitesimal automorphism.

Example 2.2. $G = \mathrm{GL}^+(n; \mathbf{R})$ and $\mathfrak{g} = \mathfrak{gl}(n; \mathbf{R})$, where $\mathrm{GL}^+(n; \mathbf{R})$ means the group of matrices with positive determinant. The Lie algebra $\mathfrak{g}$ is of infinite type. A manifold M admits a $\mathrm{GL}^+(n; \mathbf{R})$-structure if and only if it is orientable; this is more or less the definition of orientability. A $\mathrm{GL}^+(n; \mathbf{R})$-structure on M may be considered as an orientation of M and is obviously integrable. A diffeomorphism of M onto itself is an automorphism of a $GL^+(n; \mathbf{R})$-structure if and only if it is orientation preserving. Every vector field on M is an automorphism since every one-parameter group of transformations is orientation preserving.

Example 2.3. $G = SL(n; \mathbf{R})$ and $\mathfrak{g} = \mathfrak{sl}(n; \mathbf{R})$. Again, $\mathfrak{g}$ contains a matrix of rank 1 and is of infinite type. The natural action of $GL(n; \mathbf{R})$ on $\mathbf{R}^n$ induces an action of $GL(n; \mathbf{R})$ on $\Lambda^n \mathbf{R}^n$ such that

$$Av = \det(A) \cdot v \quad \text{for } A \in GL(n; \mathbf{R}) \quad \text{and } v \in \Lambda^n \mathbf{R}^n.$$

The group $GL(n; \mathbf{R})$ is transitive on $\Lambda^n \mathbf{R}^n - \{0\}$ with isotropy subgroup $SL(n; \mathbf{R})$ so that $\Lambda^n \mathbf{R}^n - \{0\} = GL(n; \mathbf{R})/SL(n; \mathbf{R})$. It follows that the cross sections of the bundle $L(M)/SL(n; \mathbf{R})$ are in one-to-one correspondence with the *volume elements* of M, i.e., the n-forms on M which vanish nowhere. In other words, an $SL(n; \mathbf{R})$-structure is nothing but a volume element on M. It is clear that M admits an $SL(n; \mathbf{R})$-structure if and only if it is orientable. We claim that every $SL(n; \mathbf{R})$-structure is integrable. Indeed, let U be a coordinate neighborhood with local coordinate system $x^1, \dots, x^n$ and let $\varphi = f\, dx^1 \wedge \dots \wedge dx^n$ be the volume element corresponding to the given $SL(n; \mathbf{R})$-structure. Let $y^1 = y^1(x^1, \dots, x^n)$ be a function such that $\partial y^1/\partial x^1 = f$. Then

$$\varphi = f\, dx^1 \wedge \dots \wedge dx^n = dy^1 \wedge dx^2 \wedge \dots \wedge dx^n,$$

which shows that the coordinate system $y^1, x^2, \dots, x^n$ is admissible with respect to the given $SL(n; \mathbf{R})$-structure. A diffeomorphism of M onto itself is an automorphism of the $SL(n; \mathbf{R})$-structure if and only if it preserves the volume element φ. Let X be a vector field on M. The function δX defined by

$$L_X \varphi = (\delta X) \cdot \varphi$$

is called the *divergence* of X with respect to φ. Clearly, X is an infinitesimal automorphism of the $SL(n; \mathbf{R})$-structure if and only if $\delta X = 0$. For $SL(n; \mathbf{R})$-structures, see § 6.

Example 2.4. $G = GL(m; \mathbf{C})$ and $\mathfrak{g} = \mathfrak{gl}(m; \mathbf{C})$. We consider $GL(m; \mathbf{C})$ (resp. $\mathfrak{gl}(m; \mathbf{C})$) as a subgroup of $GL(2m; \mathbf{R})$ (resp. a subalgebra of $\mathfrak{gl}(2m; \mathbf{R})$) in a natural manner, i.e.,

$$A_1 + i A_2 \in GL(m; \mathbf{C}) \to \begin{pmatrix} A_1 & A_2 \\ -A_2 & A_1 \end{pmatrix} \in GL(2m; \mathbf{R})$$
$$\text{or} \quad \mathfrak{gl}(m; \mathbf{C}) \qquad\qquad \text{or} \quad \mathfrak{gl}(2m; \mathbf{R}).$$

Let $z^1, \dots, z^m$ be the natural coordinate system in $\mathbf{C}^m$ and $z^j = x^j + i x^{m+j}$, $j = 1, \dots, m$. Then the identification $\mathbf{C}^m = \mathbf{R}^{2m}$ given by

$$(z^1, \dots, z^m) \to (x^1, \dots, x^{2m})$$

induces the preceding injections

$$\mathrm{GL}(m;\mathbf{C})\to\mathrm{GL}(2m;\mathbf{R})\quad\text{and}\quad\mathfrak{gl}(m;\mathbf{C})\to\mathfrak{gl}(2m;\mathbf{R}).$$

The multiplication by i in $\mathbf{C}^m$, i.e.,

$$(z^1,\ldots,z^m)\to(iz^1,\ldots,iz^m),$$

induces a linear transformation

$$(x^1,\ldots,x^m,x^{m+1},\ldots,x^{2m})\to(-x^{m+1},\ldots,-x^{2m},x^1,\ldots,x^m)$$

of $\mathbf{R}^{2m}$, which will be denoted by J. Since $i^2=-1$, we have $\mathsf{J}^2=-I$. In matrix form,

$$\mathsf{J}=\begin{pmatrix}0&-I\\I&0\end{pmatrix}.$$

The group $\mathrm{GL}(m;\mathbf{C})$ (resp. the algebra $\mathfrak{gl}(m;\mathbf{C})$), considered as a subgroup of $\mathrm{GL}(2m;\mathbf{R})$ (resp. a subalgebra of $\mathfrak{gl}(2m;\mathbf{R})$), is given by

$$\mathrm{GL}(m;\mathbf{C})=\{A\in\mathrm{GL}(2m;\mathbf{R});\ A\mathsf{J}=\mathsf{J}A\}$$
$$\mathfrak{gl}(m;\mathbf{C})=\{A\in\mathfrak{gl}(2m;\mathbf{R});\ A\mathsf{J}=\mathsf{J}A\}.$$

Since $\mathfrak{g}_k$ consists of all symmetric multilinear mappings of $\mathbf{C}^m\times\cdots\times\mathbf{C}^m$ ($k+1$ times) into $\mathbf{C}^m$, the Lie algebra $\mathfrak{g}$ is of infinite type. Every element of $\mathfrak{g}$, considered as an element of $\mathfrak{gl}(2m;\mathbf{R})$ is of even rank. Hence, $\mathfrak{g}$ is elliptic. The $\mathrm{GL}(m;\mathbf{C})$-structure on a manifold M (of dimension $2m$) are in one-to-one correspondence with the tensor field J of type (1, 1) on M such that

$$J_x\circ J_x=-I_x\qquad(\text{or simply, } J\circ J=-I),$$

where J_x is the endomorphism of the tangent space $T_x(M)$ given by J and I_x is the identity transformation of $T_x(M)$. The correspondence is given as follows. Given a tensor field J with $J\circ J=-I$, we consider, at each point x of M, only those linear frames $u\colon\mathbf{R}^{2m}\to T_x(M)$ satisfying $u\circ\mathsf{J}=J_x\circ u$. The subbundle of $L(M)$ thus obtained is the corresponding $\mathrm{GL}(m;\mathbf{C})$-structure on M. A tensor field J with $J\circ J=-I$ or the corresponding $\mathrm{GL}(m;\mathbf{C})$-structure is called an *almost complex structure*. We claim that *an almost complex structure is integrable (as a* $\mathrm{GL}(m;\mathbf{C})$*-structure) if and only if it comes from a complex structure.* (Before we explain this statement, we should perhaps remark an almost complex structure J is often called integrable if a certain tensor field of type (1, 2), called the torsion or Nijenhuis tensor, vanishes.) It is a deep result of Newlander and Nirenberg [1] that the two definitions coincide. For the real analytic case, see, for instance, Kobayashi-Nomizu [1, vol. 2; p. 321]. The theorem of Newlander-Nirenberg is equivalent to the statement that an $\mathrm{GL}(m;\mathbf{C})$-structure is integrable if and only if it admits a torsionfree affine connection (see Fröhlicher [1]). Let M be a complex manifold of complex dimension m with local coordinate system $z^1, \dots, z^m$ where $z^j = x^j + i y^j$. We have the natural almost complex structure J on M defined by

$$J(\partial/\partial x^j) = \partial/\partial y^j \qquad j=1, \dots, m,$$
$$J(\partial/\partial y^j) = -\partial/\partial x^j \quad j=1, \dots, m.$$

The almost complex structure J thus obtained is integrable since

$$(\partial/\partial x^1, \dots, \partial/\partial x^m, \partial/\partial y^1, \dots, \partial/\partial y^m)$$

gives a local cross section of the $\mathrm{GL}(m; \mathbf{C})$-structure defined by J. Conversely, if an almost complex structure J is integrable as a $\mathrm{GL}(m; \mathbf{C})$-structure and if $x^1, \dots, x^{2m}$ is an admissible local coordinate system, then $J(\partial/\partial x^j) = \partial/\partial x^{m+j}$ and $J(\partial/\partial x^{m+j}) = -\partial/\partial x^j$ for $j=1, \dots, m$. If we set $z^j = x^j + i x^{m+j}$, then the complex coordinate system $z^1, \dots, z^m$ turns M into a complex manifold. A diffeomorphism f of M onto itself is an automorphism of an almost complex structure J if and only if $f_* \circ J = J \circ f_*$, where $f_*\colon T(M) \to T(M)$ is the differential of f. If J is integrable, an automorphism f is nothing but a holomorphic diffeomorphism. A vector field X on M is an infinitesimal automorphism of an almost complex structure J if and only if

$$[X, JY] = J([X, Y]) \qquad \text{for all vector field } Y \text{ on } M.$$

For further properties of an almost complex structure, see Kobayashi-Nomizu [1; Chapter IX].

Example 2.5. $G = \mathrm{O}(n)$ and $\mathfrak{g} = \mathfrak{o}(n)$. The Lie algebra $\mathfrak{g}$ is of order 1. Let $t \in \mathfrak{g}_1$ and (t^i_{jk}) the components of t. By definition, $t^i_{jk} = t^i_{kj}$. Since $\mathfrak{o}(n)$ consists of skew-symmetric matrices, we have $t^i_{jk} = -t^j_{ik}$. Hence,

$$t^i_{jk} = -t^j_{ik} = -t^j_{ki} = t^k_{ji} = t^k_{ij} = -t^i_{kj} = -t^i_{jk},$$

thus proving $t^i_{jk} = 0$. To each Riemannian metric on M, there corresponds the bundle of orthonormal frames over M. This gives a one-to-one correspondence between the Riemannian metrics on M and the $\mathrm{O}(n)$-structures on M. An $\mathrm{O}(n)$-structure is integrable if and only if the corresponding Riemannian metric is flat, i.e., it has vanishing curvature. An automorphism of an $\mathrm{O}(n)$-structure is an isometry of the corresponding Riemannian metric. An infinitesimal automorphism of an $\mathrm{O}(n)$-structure is an infinitesimal isometry or Killing vector field. We shall discuss isometries and Killing vector fields in detail later (see Chapter II).

More generally, let $G = \mathrm{O}(p, q)$, $n = p + q$, be the orthogonal group defined by a quadratic form $u_1^2 + \dots + u_p^2 - u_{p+1}^2 - \dots - u_n^2$. Then $\mathfrak{o}(p, q)$ is also of order 1. There is a natural one-to-one correspondence between the pseudo-Riemannian metrics of signature q on M and the $O(p,q)$-structures on M. An $O(p,q)$-structure is integrable if and only if the corresponding pseudo-Riemannian metric has vanishing curvature. It should be remarked that, although every paracompact manifold admits a Riemannian metric, it may not in general admit a pseudo-Riemannian metric of signature q for $q \neq 0, n$. For automorphism of pseudo-Riemannian manifolds, see Tanno [1, 2].

Example 2.6. $G = CO(n)$ and $\mathfrak{g} = \mathfrak{co}(n)$, $n \geqq 3$. By definition,

$$CO(n) = \{A \in GL(n; \mathbf{R});\ {}^tAA = cI,\ c \in \mathbf{R},\ c > 0\},$$

$$\mathfrak{co}(n) = \{A \in \mathfrak{gl}(n; \mathbf{R});\ {}^tA + A = cI,\ c \in \mathbf{R}\}.$$

Thus, $CO(n) = O(n) \times \mathbf{R}^+$ and $\mathfrak{co}(n) = \mathfrak{o}(n) + \mathbf{R}$, where $\mathbf{R}^+$ denotes the multiplicative group of positive real numbers. The Lie algebra $\mathfrak{co}(n)$ is of order 2 and the first prolongation $\mathfrak{g}_1$ is naturally isomorphic to the dual space $\mathbf{R}^{n*}$ of $\mathbf{R}^n$. To determine $\mathfrak{g}_1$, let $t = (t^i_{jk})$ be an element of $\mathfrak{g}_1$. Since the kernel of the homomorphism $A \in \mathfrak{co}(n) \to \operatorname{trace}(A) \in \mathbf{R}$ is precisely $\mathfrak{o}(n)$ and since $\mathfrak{o}(n)$ is of order 1, the linear mapping

$$t = (t^i_{jk}) \in \mathfrak{g}_1 \to \xi = \left(\frac{1}{n} \sum_i t^i_{ik}\right) \in \mathbf{R}^{n*}$$

is injective. The kernel is the first prolongation of $\mathfrak{o}(n)$. (The factor of $\frac{1}{n}$ is, of course, not important). To see that this mapping is also surjective, we have only to observe that $\xi = (\xi_k)$ is the image of t with $t^i_{jk} = \delta^i_j \xi_k + \delta^i_k \xi_j - \delta^j_k \xi_i$. To prove $\mathfrak{g}_2 = 0$, let $t = (t^h_{ijk}) \in \mathfrak{g}_2$. For each fixed k, t^h_{ijk} may be considered as the components of an element in $\mathfrak{g}_1$ and hence can be uniquely written

$$t^h_{ijk} = \delta^h_i \xi_{jk} + \delta^h_j \xi_{ik} - \delta^i_j \xi_{hk}.$$

Since t^h_{ijk} must be symmetric in all lower indices, we have $\sum_h t^h_{hjk} = \sum_h t^h_{hkj}$, from which follows $\xi_{jk} = \xi_{kj}$. From $\sum_h t^h_{hjk} = \sum_h t^h_{jkh}$, we obtain $(n-2)\xi_{jk} = -\delta_{jk} \cdot \sum_h \xi_{hh}$, from which follows $(n-2)\sum_h \xi_{hh} = -n \sum_h \xi_{hh}$ and, hence, $\sum_h \xi_{hh} = 0$. From $(n-2)\xi_{jk} = -\delta_{jk} \cdot \sum_h \xi_{hh} = 0$ and $n \geqq 3$, we conclude $\xi_{jk} = 0$. (The reader who prefers an index-free proof is referred to Kobayashi-Nagano [3, III; p. 686].) A $CO(n)$-structure is called a *conformal structure*. We say that two Riemannian metrics on M are *conformally equivalent* if one is a multiple of the other by a positive function. The conformal equivalence classes of Riemannian metrics on M are in a natural one-to-one correspondence with the $CO(n)$-structure on M. A conformal structure is integrable if and only if any Riemannian metric corresponding to the structure is locally conformally equivalent to $(dx^1)^2+\dots+(dx^n)^2$ with respect to a suitable local coordinate system $x^1, \dots, x^n$. Thus, a conformal structure is integrable if and only if it is conformally flat in the classical sense (see Eisenhart [1]). Consequently, the integrability of a conformal structure is equivalent to the vanishing of the so-called conformal curvature tensor of Weyl (provided $n \geqq 3$). Given a Riemannian metric g on M, a diffeomorphism f of M onto itself (resp. a vector field X on M) is a conformal transformation, i.e., an automorphism of the conformal structure (resp. an infinitesimal conformal transformation, i.e., an infinitesimal automorphism of the conformal structure) if and only if

$$f^* g = \rho \cdot g \qquad (\text{resp. } L_X g = \sigma \cdot g),$$

where ρ (resp. σ) is a positive function (resp. a function) on M. Conformal structures and their automorphisms will be discussed in Chapter IV.

The reason we excluded the case $n=2$ is that $CO(2)$ (resp. $\mathfrak{co}(2)$) is naturally isomorphic to $GL(1; \mathbf{C})$ (resp. $\mathfrak{gl}(1; \mathbf{C})$). For this reason, the conformal differential geometry in dimension 2 differs significantly from that in higher dimensions. In particular, we note that every $CO(2)$-structure, i.e., $GL(1; \mathbf{C})$-structure is integrable; this is nothing but the existence of isothermal coordinate systems.

The results for $CO(n)$-structures can be easily generalized to $CO(p, q)$-structures, where $CO(p, q) = O(p, q) \times \mathbf{R}^+$ is defined by a quadratic form of signature q.

Example 2.7. $G = U(m)$ and $\mathfrak{g} = \mathfrak{u}(m)$. Since $\mathfrak{u}(m)$ is a subalgebra of $\mathfrak{o}(2m)$ which is of order 1 (cf. Examples 2.4 and 2.5), it is also of order 1. A $U(m)$-structure on a $2m$-dimensional manifold M is called an *almost hermitian structure;* it consists of an almost complex structure and a hermitian metric. Since $U(m) = GL(m; \mathbf{C}) \cap O(2m)$, a $U(m)$-structure may be considered as an intersection of a $GL(m; \mathbf{C})$-structure and an $O(2m)$-structure. A $U(m)$-structure is integrable if and only if the underlying almost complex structure is integrable (so that M is a complex manifold) and the hermitian metric has vanishing torsion and curvature. A diffeomorphism of M onto itself is an automorphism of a $U(m)$-structure if and only if it is an automorphism of the underlying $GL(m; \mathbf{C})$- and $O(2m)$-structures. Similarly, for an infinitesimal automorphism. For automorphisms of hermitian manifolds, see Tanno [3].

Example 2.8. $G = Sp(m; \mathbf{R})$ and $\mathfrak{g} = \mathfrak{sp}(m; \mathbf{R})$. We recall that $Sp(m; \mathbf{R})$ is the group of linear transformations of $\mathbf{R}^{2m}$ leaving the form

$$u^1 \wedge u^{m+1} + \dots + u^m \wedge u^{2m}$$

invariant, where $u^1, \ldots, u^{2m}$ is the natural coordinate system in $\mathbf{R}^{2m}$. In other words,

$$\mathrm{Sp}(m; \mathbf{R}) = \{A \in \mathrm{GL}(2m; \mathbf{R});\ {}^tA\mathsf{J}A = \mathsf{J}\},$$

$$\mathfrak{sp}(m; \mathbf{R}) = \{A \in \mathfrak{gl}(2m; \mathbf{R});\ {}^tA\mathsf{J} + \mathsf{J}A = 0\},$$

where

$$\mathsf{J} = \begin{pmatrix} 0 & -I \\ I & 0 \end{pmatrix}.$$

Since $\mathfrak{sp}(m; \mathbf{R})$ consists of matrices of the form

$$A = \begin{pmatrix} A_1 & A_2 \\ A_3 & {}^tA_1 \end{pmatrix} \quad \text{with } {}^tA_2 = A_2 \quad \text{and} \quad {}^tA_3 = A_3,$$

it contains an element of rank 1 and, hence, is of infinite type. The $\mathrm{Sp}(m, \mathbf{R})$-structures on a $2m$-dimensional manifold M are in a natural one-to-one correspondence with the 2-forms ω on M of maximum rank (i.e., $\omega^m \neq 0$ everywhere).

Since both $\mathrm{GL}(m; \mathbf{C})$ and $\mathrm{Sp}(m; \mathbf{R})$ contain $U(m)$ as a maximal compact subgroup, a manifold M admits an $\mathrm{Sp}(m; \mathbf{R})$-structure if and only if it admits a $\mathrm{GL}(m; \mathbf{C})$-structure. An $\mathrm{Sp}(m; \mathbf{R})$-structure is called an *almost symplectic structure* or an *almost Hamiltonian structure.* If an almost symplectic structure is integrable with admissible coordinate system $x^1, \ldots, x^{2m}$ so that

$$\omega = dx^1 \wedge dx^{m+1} + \cdots + dx^m \wedge dx^{2m},$$

then $d\omega = 0$. Conversely (see Appendix 1), if the form ω defining an almost symplectic structure is closed, then $\omega = dx^1 \wedge dx^{m+1} + \cdots + dx^m \wedge dx^{2m}$ for a suitable local coordinate system $x^1, \ldots, x^{2m}$ and the structure is integrable. An integrable almost symplectic structure is called a *symplectic structure* or a *Hamiltonian structure.* We observe that if an almost symplectic structure admits a torsionfree affine connection, then it is integrable. For the 2-form ω defining an almost symplectic structure is parallel with respect to such a connection and hence is closed. (In calculating $d\omega$ in terms of a local coordinate system, partial differentiation may be replaced by covariant differentiation when the connection is torsionfree, see for instance Kobayashi-Nomizu [1, vol. 1; p. 149]). A diffeomorphism f of M onto itself is an automorphism of the symplectic structure defined by a 2-form ω if and only if $f^*\omega = \omega$. Similarly, X is an infinitesimal automorphism if and only if $L_X\omega = 0$. An (infinitesimal) automorphism of a symplectic structure is called an *(infinitesimal) symplectic transformation.*

Set

$$\mathrm{CSp}(m; \mathbf{R}) = \{A \in \mathrm{GL}(2m; \mathbf{R});\ {}^tA\mathsf{J}A = c\mathsf{J},\ c \in \mathbf{R}^+\} = \mathrm{Sp}(m; \mathbf{R}) \times \mathbf{R}^+,$$

$$\mathfrak{csp}(m; \mathbf{R}) = \{A \in \mathfrak{gl}(2m; \mathbf{R});\ {}^tA\mathsf{J} + \mathsf{J}A = c\mathsf{J},\ c \in \mathbf{R}\} = \mathfrak{sp}(m; \mathbf{R}) + \mathbf{R}.$$

A $\mathrm{CSp}(m; \mathbf{R})$-structure is called a *conformal-symplectic structure*. For conformal-symplectic geometry, see Lee [1].

Example 2.9. $G = \mathrm{GL}(p, q; \mathbf{R})$ and $\mathfrak{g} = \mathfrak{gl}(p, q; \mathbf{R})$, where $\mathrm{GL}(p, q; \mathbf{R})$ denotes the group of linear transformations of $\mathbf{R}^n$, $n = p + q$, which leave the p-dimensional subspace $\mathbf{R}^p$ defined by $u^{p+1} = \cdots = u^n = 0$ invariant. In other words,

$$\mathrm{GL}(p, q; \mathbf{R}) = \left\{ \begin{pmatrix} A & B \\ 0 & C \end{pmatrix}; \ A \in \mathrm{GL}(p; \mathbf{R}),\ C \in \mathrm{GL}(q; \mathbf{R}) \right\}$$

$$\mathfrak{gl}(p, q; \mathbf{R}) = \left\{ \begin{pmatrix} A & B \\ 0 & C \end{pmatrix}; \ A \in \mathfrak{gl}(p; \mathbf{R}),\ C \in \mathfrak{gl}(q; \mathbf{R}) \right\},$$

where B denotes a matrix with p rows and q columns. Clearly, $\mathfrak{g}$ contains an element of rank 1 and, hence, is of infinite type. The $\mathrm{GL}(p, q; \mathbf{R})$-structures on M are in a natural one-to-one correspondence with the p-dimensional distributions on M, i.e., the fields of p-dimensional subspaces of tangent spaces. A $\mathrm{GL}(p, q; \mathbf{R})$-structure is integrable if and only if there exists a local coordinate system $x^1, \ldots, x^n$ such that $\partial/\partial x^1, \ldots, \partial/\partial x^p$ span the corresponding p-dimensional distribution. In other words, a $\mathrm{GL}(p, q; \mathbf{R})$-structure is integrable if and only if the corresponding p-dimensional distribution is involutive, (see Frobenius theorem). An integrable $\mathrm{GL}(p, q; \mathbf{R})$-structure is known as a *foliation* with p-dimensional leaves. If a $\mathrm{GL}(p, q; \mathbf{R})$-structure admits a torsionfree affine connection, it is integrable. Indeed, if X and Y are vector fields belonging to the distribution, then the formula $[X, Y] = \nabla_X Y - \nabla_Y X$ (see Kobayashi-Nomizu [1; p. 133]) implies that $[X, Y]$ also belongs to the distribution. Since an automorphism of a $\mathrm{GL}(p, q; \mathbf{R})$-structure on M is a transformation preserving the corresponding p-dimensional distribution, a vector field X on M is an infinitesimal automorphism if and only if, for every vector field Y belonging to the distribution, $[X, Y]$ belongs to the distribution.

Example 2.10. $G = \mathrm{GL}(p; \mathbf{R}) \times \mathrm{GL}(q; \mathbf{R})$ and $\mathfrak{g} = \mathfrak{gl}(p; \mathbf{R}) + \mathfrak{gl}(q; \mathbf{R})$, $p + q = n$. In other words,

$$\mathrm{GL}(p; \mathbf{R}) \times \mathrm{GL}(q; \mathbf{R}) = \left\{ \begin{pmatrix} A & 0 \\ 0 & B \end{pmatrix}; \ A \in \mathrm{GL}(p; \mathbf{R}),\ B \in \mathrm{GL}(q; \mathbf{R}) \right\},$$

$$\mathfrak{gl}(p; \mathbf{R}) + \mathfrak{gl}(q; \mathbf{R}) = \left\{ \begin{pmatrix} A & 0 \\ 0 & B \end{pmatrix}; \ A \in \mathfrak{gl}(p; \mathbf{R}),\ B \in \mathfrak{gl}(q; \mathbf{R}) \right\}.$$

Clearly, $\mathfrak{g}$ contains an element of rank 1 and, hence, is of infinite type. The $\mathrm{GL}(p; \mathbf{R}) \times \mathrm{GL}(q; \mathbf{R})$-structures are in a natural one-to-one correspondence with the set of pairs (S, S'), where S and S' are complementary distributions of dimensions p and q respectively. A $\mathrm{GL}(p; \mathbf{R}) \times \mathrm{GL}(q; \mathbf{R})$-

structure is integrable if and only if the corresponding distributions S and S' are both involutive, that is, there exists a local coordinate system $x^1, \ldots, x^n$ such that $\partial/\partial x^1, \ldots, \partial/\partial x^p$ span S and $\partial/\partial x^{p+1}, \ldots, \partial/\partial x^n$ span S'.

Example 2.11. $G=\{1\}$ and $\mathfrak{g}=0$. The $\{1\}$-structures on M are in a natural one-to-one correspondence with the fields of linear frames over M. A manifold M is said to be *parallelisable* if it admits a $\{1\}$-structure. The automorphism group of a $\{1\}$-structure will be studied in the next section (Theorem 3.2).

3. Two Theorems on Differentiable Transformation Groups

The theorems in this section will allow us to prove that the automorphism groups of many geometric structures are Lie groups.

Theorem 3.1. *Let $\mathfrak{G}$ be a group of differentiable transformations of a manifold M. Let S be the set of all vector fields X on M which generate global* 1-*parameter groups $\varphi_t=\exp tX$ of transformations of M such that $\varphi_t\in\mathfrak{G}$. If the set S generates a finite-dimensional Lie algebra of vector fields on M, then $\mathfrak{G}$ is a Lie transformation group and S is the Lie algebra of $\mathfrak{G}$.*

Proof. Let $\mathfrak{g}^*$ be the Lie algebra of vector fields on M generated by S. Let $\tilde{\mathfrak{G}}$ be the connected, simply connected Lie group with Lie algebra $\mathfrak{g}^*$; it is an abstract Lie group and is not a transformation group. For each element X of $\mathfrak{g}^*$, we denote by e^{tX} the 1-parameter subgroup of $\tilde{\mathfrak{G}}$ generated by X while we denote by $\exp tX$ the 1-parameter local group of local transformations of M generated by the vector field X. Then the group $\tilde{\mathfrak{G}}$ acts *locally* on M in the following sense. There exist a neighborhood U of $\{1\}\times M$ in $\tilde{\mathfrak{G}}\times M$ and a mapping $f\colon U\to M$ such that

$$f(e^{tX}, p)=(\exp tX)\,p \qquad \text{for } (e^{tX}, p)\in U\subset\tilde{\mathfrak{G}}\times M.$$

Lemma 1. *Given X, $Y\in\mathfrak{g}^*$, we define $Z\in\mathfrak{g}^*$ by $e^{tZ}=e^X e^{tY} e^{-X}$,* i.e., $Z=(\operatorname{ad} e^X)\,Y$. *If X, Y are in S, so is Z.*

Proof of Lemma 1. From $e^{tZ}=e^X e^{tY} e^{-X}$, we obtain

$$(\exp tZ)\,p=(\exp X)(\exp tY)(\exp -X)\,p.$$

If X, $Y\in S$, then the right hand side is defined for all p and t. Hence, $(\exp tZ)\,p$ is also defined for all p and t. This implies that Z is in S.

Lemma 2. S spans $\mathfrak{g}^*$ *as a vector space.*

Proof of Lemma 2. Let V be the vector subspace of $\mathfrak{g}^*$ spanned by S. By Lemma 1, we have $(\operatorname{ad} e^S)\,S\subset S$ and, hence, $(\operatorname{ad} e^S)\,V\subset V$. Since S

generates $\mathfrak{g}^*$, e^S generates $\mathfrak{G}$. Hence, $(\operatorname{ad}\mathfrak{G})\,V\subset V$. In particular, $(\operatorname{ad} e^V)\cdot V\subset V$, which implies $[V, V]\subset V$.

Lemma 3. $S=\mathfrak{g}^*$.

Proof of Lemma 3. Let $X_1, \ldots, X_r\in S$ be a basis for $\mathfrak{g}^*$. Then the mapping

$$\sum a_i X_i\in\mathfrak{g}^*\to e^{a_1 X_1}\ldots e^{a_r X_r}\in\mathfrak{G}$$

gives a diffeomorphism of a neighborhood N of 0 in $\mathfrak{g}^*$ onto a neighborhood U of the identity element in $\mathfrak{G}$. Let $Y\in\mathfrak{g}^*$. Let δ be a positive number such that $e^{tY}\in U$ for $|t|<\delta$. Then, for each t with $|t|<\delta$, there exists a unique element $\sum a_i(t)\,X_i\in N$ such that

$$e^{tY}=e^{a_1(t)X_1}\ldots e^{a_r(t)X_r}.$$

The action of $\exp tY$ on M is therefore given by

$$(\exp tY)\,p=(\exp a_1(t)\,X_1)\ldots(\exp a_r(t)\,X_r)\,p \quad \text{for } p\in M \quad \text{and} \quad |t|<\delta.$$

This shows that every element Y of $\mathfrak{g}^*$ generates a global 1-parameter group of transformations of M. Hence, $Y\in S$, thus completing the proof of Lemma 3.

Let $\mathfrak{G}^*$ be the Lie transformation group acting on M generated by $\mathfrak{g}^*$; $\mathfrak{G}^*$ exists since every element of $\mathfrak{g}^*$ generates a global 1-parameter group of transformations of M by Lemma 3. Since $\mathfrak{G}^*$ is connected, the assumption in the statement of Theorem 3.1 implies $\mathfrak{G}^*\subset\mathfrak{G}$. Let $\varphi\in\mathfrak{G}$ and ψ_t be a 1-parameter subgroup of $\mathfrak{G}^*$. Then $\varphi\,\psi_t\,\varphi^{-1}$ is a 1-parameter group of transformations of M contained in $\mathfrak{G}$. From the contruction of $\mathfrak{G}^*$ it follows that this 1-parameter group is a subgroup of $\mathfrak{G}^*$. Since $\mathfrak{G}^*$ is generated by its 1-parameter subgroups, this implies that $\mathfrak{G}^*$ is a normal subgroup of $\mathfrak{G}$. Each $\varphi\in\mathfrak{G}$ defines an automorphism A_φ: $\mathfrak{G}^*\to\mathfrak{G}^*$ by $A_\varphi(\psi)=\varphi\,\psi\,\varphi^{-1}$. Since the automorphism A_φ sends every 1-parameter subgroup of $\mathfrak{G}^*$ into a 1-parameter subgroup of $\mathfrak{G}^*$, it is continuous (see Chevalley [1; p. 128]).

Lemma 4. *Let $\mathfrak{G}$ be a group and $\mathfrak{G}^*$ a topological group contained in $\mathfrak{G}$ as a normal subgroup. If A_φ: $\mathfrak{G}^*\to\mathfrak{G}^*$ is continuous for each $\varphi\in\mathfrak{G}$, then there exists a unique topology on $\mathfrak{G}$ which makes $\mathfrak{G}^*$ open in $\mathfrak{G}$.*

Proof of Lemma 4. If $\{V\}$ is the system of open neighborhoods of the identity element in $\mathfrak{G}^*$, we take $\{\varphi(V)\}$ as the system of open neighborhoods of $\varphi\in\mathfrak{G}$ in $\mathfrak{G}$. It is a trivial matter to verify that $\mathfrak{G}^*$ is open in $\mathfrak{G}$ with respect to the topology thus defined in G. The uniqueness of such a topology is also evident.

Applying Lemma 4 to our case, we see that $\mathfrak{G}$ is a topological group with identity component $\mathfrak{G}^*$. Since $\mathfrak{G}^*$ is a Lie group, its differentiable structure can be translated to other connected components of $\mathfrak{G}$. The differentiability of the action $\mathfrak{G}^* \times M \to M$ implies that of $\mathfrak{G} \times M \to M$. q.e.d.

As we shall see in § 5 (Theorem 5.1) the study of the automorphism group of a G-structure can be reduced to the case of a $\{1\}$-structure if the Lie algebra $\mathfrak{g}$ is of finite type. The following theorem is therefore basic.

Theorem 3.2. *Let M be a manifold with a $\{1\}$-structure (i.e., an absolute parallelism). Let $\mathfrak{A}$ be the group of automorphisms of the $\{1\}$-structure. Then $\mathfrak{A}$ is a Lie transformation group such that* $\dim \mathfrak{A} \leq \dim M$. *More precisely, for any point $p \in M$, the mapping $a \in \mathfrak{A} \to a(p) \in M$ is injective and its image $\{a(p);\, a \in \mathfrak{A}\}$ is a closed submanifold of M. The submanifold structure on this image makes $\mathfrak{A}$ into a Lie transformation group.*

Proof. Let $e_1, \ldots, e_n$ be everywhere linearly independent vector fields on M defining the given $\{1\}$-structure. Let V be the set of vector fields v which are linear combinations of $e_1, \ldots, e_n$ with constant coefficients. Then V is a vector space of dimension n. By definition, $\mathfrak{A}$ consists of transformations a of M which leave each $v \in V$ invariant. In other words,

$$(*) \qquad a \circ \exp v = (\exp v) \circ a \qquad \text{for } a \in \mathfrak{A},\ v \in V,$$

wherever $\exp v$ is defined. (In general, $(\exp tv)\,p$ is defined only for small values of t depending upon the point $p \in M$.)

Lemma 1. *The mapping $a \in \mathfrak{A} \to a(p) \in M$ is injective.*

Proof of Lemma 1. Let F_a be the fixed point set of $a \in \mathfrak{A}$. It is a closed subset of M. If $q \in F_a$, then the set of points $(\exp v)\,q$ covers a neighborhood of q when v varies in a neighborhood of the origin in V. Hence, the equality $(*)$ implies that this neighborhood of q is in F_a. Since F_a is closed and open, either $F_a = M$ so that a is the identity element or F_a is empty.

Lemma 2. *The set $\{a(p);\, a \in \mathfrak{A}\}$ is closed in M for each $p \in M$.*

Proof of Lemma 2. Let $\{a_k\}$ be a sequence of elements of $\mathfrak{A}$ such that $a_k(p) \to q$ for some $q \in M$. We want to construct an element a of $\mathfrak{A}$ such that $a(p) = q$. We define the transformation a first in a neighborhood of p by setting

$$a((\exp v)\,p) = (\exp v)\,q$$

for all v which are in a neighborhood of the origin in V so that both $(\exp v)\,p$ and $(\exp v)\,q$ are defined. Then $a(p') = \lim a_k(p')$ for all p' in a neighborhood of p for which the transformation a is thus defined. Using $(*)$ we extend the definition of a along curves from p. Since $a(p') = \lim a_k(p')$, the extended map a is independent of the choice of curves. From the construction of a, it is clear that a is a local diffeomorphism. To see that a^{-1} exists, we observe first that $a_k^{-1}(q) \to p$ and then apply the same construction to obtain a^{-1} as the limits of $\{a_k^{-1}\}$. It is easy to see that a is an automorphism of the $\{1\}$-structure.

Let $\mathfrak{l}$ be the set of vector fields X on M such that $[X, v]=0$ for all v in V. It is a Lie algebra of vector fields.

Lemma 3. *For each point $p \in M$, the restriction map $X \in \mathfrak{l} \to X_p \in T_p(M)$ is injective. In particular,* $\dim \mathfrak{l} \leqq \dim M$.

Proof of Lemma 3. This is immediate from (*).

Let $\mathfrak{a}$ be the set of vector fields $X \in \mathfrak{l}$ which generate a global 1-parameter group of transformations of M. The Lie algebra of vector fields generated by the set $\mathfrak{a}$ is contained in $\mathfrak{l}$ and hence is of finite dimension. We apply Theorem 3.1 to $\mathfrak{G}=\mathfrak{A}$ and $S=\mathfrak{a}$. Then we can conclude that $\mathfrak{a}$ is a Lie algebra and $\mathfrak{A}$ is a Lie transformation group with Lie algebra $\mathfrak{a}$. Since the action $\mathfrak{A} \times M \to M$ is differentiable and since the image of the injection $a \in \mathfrak{A} \to a(p) \in M$ is closed, $\{a(p);\, a \in \mathfrak{A}\}$ is a closed submanifold of M and the mapping

$$a \in \mathfrak{A} \to a(p) \in \{a(p);\, a \in \mathfrak{A}\}$$

is a diffeomorphism. q.e.d.

Theorem 3.1 is due to Palais [1]; the proof given here is from Chu-Kobayashi [1]. Theorem 3.2 is due to Kobayashi [1]; the original proof was closer to that of Myers-Steenrod [1] for the group of isometries of a Riemannian manifold.

Another important criterion for a topological transformation group to be a Lie transformation group is the following theorem of Bochner-Montgomery [1] we state without proof (see also Montgomery-Zippin [1]).

Theorem 3.3. *Let $\mathfrak{G}$ be a locally compact group of differentiable transformations of a manifold M. Then $\mathfrak{G}$ is a Lie transformation group.*

4. Automorphisms of Compact Elliptic Structures

We recall that a linear Lie algebra $\mathfrak{g} \subset \mathfrak{gl}(n; \mathbf{R})$ is said to be elliptic if $\mathfrak{g}$ contains no matrix of rank 1 (see § 1) and that $\mathfrak{gl}(m; \mathbf{C})$, as a subalgebra of $\mathfrak{gl}(2m; \mathbf{R})$, is elliptic (see Example 2.4). The purpose of this section is to prove

Theorem 4.1. *Let P be a G-structure on an n-dimensional compact manifold M. If $\mathfrak{g}$ is elliptic, then the group of automorphisms of P is a Lie transformation group.*

Proof. We shall show that the Lie algebra of infinitesimal automorphisms of P is finite dimensional. Then the theorem will follow directly from Theorem 3.1. (We remark that the proof of Theorem 3.1 becomes extremely simple when M is compact because every vector field on M generates a global 1-parameter group of transformations.)

The essential idea is to construct a system of elliptic partial differential equations of which the infinitesimal automorphisms of P are solutions.

Since $\mathfrak{g}$ is a linear subspace of $\mathfrak{gl}(n; \mathbf{R})$, it may be defined by

$$\mathfrak{g} = \left\{ (a^i_j) \in \mathfrak{gl}(n; \mathbf{R});\ \sum_{i,j=1}^{n} c^j_{i\lambda}\, a^i_j = 0 \text{ for } \lambda = 1, \ldots, N \right\},$$

where the $c^j_{i\lambda}$ are constants and N is the codimension of $\mathfrak{g}$ in $\mathfrak{gl}(n; \mathbf{R})$.

Let $V_1, \ldots, V_n$ be vector fields locally defined on M which define a local cross section of P. Let $\omega^1, \ldots, \omega^n$ be the dual basis of $V_1, \ldots, V_n$; they are linearly independent 1-forms such that $\omega^i(V_j) = \delta^i_j$. Let X be an infinitesimal automorphism of P and write

$$X = \sum \xi^i\, V_i.$$

Fixing a connection in P, denote by ∇ its covariant differentiation operator. Then write

$$\nabla X = \sum_{i,j} \xi^i{}_{;j}\, \omega^j \otimes V_i.$$

(The coefficients $\xi^i{}_{;j}$ are defined by the equality above.) Let $x^1, \ldots, x^n$ be a local coordinate system in M. Then

$$(1) \qquad \xi^i{}_{;j} = \sum \frac{\partial \xi^i}{\partial x^k} A^k_j(x) + \cdots,$$

where (A^k_j) is defined by $\sum A^k_j\, \omega^j = d x^k$ and the dots indicate the terms not involving partial derivatives of X^i.

In the definition of the torsion tensor

$$T(X, V_i) = \nabla_X V_i - \nabla_{V_i} X - [X, V_i],$$

we have

$$(2) \qquad \nabla_X V_i = \sum \mu^j_i(x)\, V_j \quad \text{with } (\mu^j_i(x)) \in \mathfrak{g}$$

and

$$(3) \qquad [X, V_i] = \sum \nu^j_i(x)\, V_j \quad \text{with } (\nu^j_i(x)) \in \mathfrak{g}.$$

While (2) follows from the fact that P is invariant under parallel displacement and $V_1, \ldots, V_n$ is a frame field belonging to P, (3) follows from the fact that $\exp(tX)$ is a 1-parameter group of automorphisms of P. On the other hand, we can write

(4) $$\nabla_{V_i} X = \sum \xi^j{}_{;i} V_j$$

and

(5) $$T(X, V_i) = \sum \tfrac{1}{2} T^k_{ji} \xi^j V_k$$

where T^k_{ji} are defined by

$$T(V_j, V_i) = \sum \tfrac{1}{2} T^k_{ji} V_k.$$

From

$$\nabla_{V_i} X + T(X, V_i) = \nabla_X V_i - [X, V_i],$$

we obtain

(6) $$\xi^j{}_{;i} + \tfrac{1}{2} \sum T^j_{ki} \xi^k = \mu^j_i - \nu^j_i.$$

Since $(\mu^j_i - \nu^j_i)$ belongs to the Lie algebra $\mathfrak{g}$, (6) implies

$$\sum c^i_{j\lambda} (\xi^j{}_{;i} + \tfrac{1}{2} \sum T^j_{ki} \xi^k) = 0 \quad \text{for } \lambda = 1, \ldots, N.$$

Substituting (1) into this, we obtain

(7) $$\sum b^k_{j\lambda} \frac{\partial \xi^j}{\partial x^k} + \cdots = 0 \quad \text{with } b^k_{j\lambda} = \sum c^i_{j\lambda} A^k_i,$$

where the dots indicate terms not involving partial derivatives of ξ^j. Differentiating (7) with respect to x^h and multiplying by $b^h_{m\lambda}$, we obtain

(8) $$\sum_{h,j,k,\lambda} b^h_{m\lambda} b^k_{j\lambda} \frac{\partial^2 \xi^j}{\partial x^k \partial x^h} + \cdots = 0,$$

where the dots indicate terms involving partial derivatives of order 1 or less. We shall show that (8) is a system of elliptic partial differential equations. We consider the following symbol of (8), which is an $n \times n$ matrix:

$$\Big(\sum_{h,k,\lambda} b^h_{m\lambda} b^k_{j\lambda} v_k v_h \Big)_{j,m=1,\ldots,n},$$

where $v = (v_1, \ldots, v_n)$ is an arbitrary nonzero covector. The problem is to show that this matrix is non-singular. Let $s = (s^1, \ldots, s^n)$ be a vector. We want to show that if

(9) $$\sum_{h,j,k,\lambda} b^h_{m\lambda} b^k_{j\lambda} v_k v_h s^j = 0,$$

then $s = 0$. Multiply (9) by s^m. Summing over the index m, we obtain

(10) $$\sum t^2_\lambda = 0, \quad \text{where } t_\lambda = \sum_{j,k} b^k_{j\lambda} v_k s^j.$$

Hence, $t_\lambda = 0$, i.e.,

$$\sum_{i,j,k} c^i_{j\lambda} A^k_i v_k s^j = 0 \qquad \text{for } \lambda = 1, \ldots, N. \tag{11}$$

Hence, the matrix $(\sum_k A^k_i v_k s^j)_{i,j=1,\ldots,n}$ belongs to the Lie algebra $\mathfrak{g}$. This matrix is the product of two matrices (A^k_i) and $(v_k s^j)$. Since (A^k_i) is non-singular and $(v_k s^j)$ is of rank 1 if $s \neq 0$ and since $\mathfrak{g}$ contains no matrix of rank 1, it follows that $s = 0$.

We have established that the infinitesimal automorphisms X of P satisfy a system of elliptic partial differential equations (of second order) (8). It follows (see, for instance, Bochner [3]) that the Lie algebra of infinitesimal automorphisms X is finite dimensional. (See also Douglis-Nirenberg [1]. If we choose any Riemannian metric on M and denote by X', X'', X''' the first, second and third covariant derivatives of X, then the Lie algebra of infinitesimal automorphisms X of P is a Banach space with norm $\|X\|$ defined by

$$\|X\| = \operatorname*{Max}_{p\in M} |X| + \operatorname*{Max}_{p\in M} |X'| + \operatorname*{Max}_{p\in M} |X''| + \operatorname*{Max}_{p\in M} |X'''|.$$

From a theorem of Douglis and Nirenberg, it follows that this Banach space is locally compact and hence is finite dimensional, (see Ruh [1] for more details)). q.e.d.

Corollary 4.2. *The automorphism group of a compact almost complex manifold is a Lie transformation group.*

Corollary 4.2 was proved by Boothby-Kobayashi-Wang [1] in the same manner as Theorem 4.1. Its generalization, Theorem 4.1, is due to Ochiai [2]. In the locally flat case, Theorem 4.1 was proved by Guillemin-Sternberg [1]. Ruh [1] also proved a similar result. As in Ochiai [2], Theorem 4.1 can be proved without the aid of a connection in P.

5. Prolongations of G-Structures

Let $V = \mathbf{R}^n$ and G be a Lie group of linear transformations of V. We recall (see § 1) that the first prolongation $\mathfrak{g}_1$ of the Lie algebra $\mathfrak{g}$ of G is the space of symmetric bilinear mappings $t\colon V \times V \to V$ such that, for each fixed $v_1 \in V$, the mapping $v \in V \to t(v, v_1) \in V$ is in $\mathfrak{g}$. We define now the *first prolongation* G_1 of G to be the group consisting of those linear transformations $\bar{t}$ of $V + \mathfrak{g}$ induced by the elements t of $\mathfrak{g}_1$ as follows:

$$\bar{t}(v) = v + t(\cdot, v) \qquad \text{for } v \in V,$$

$$\bar{t}(x) = x \qquad \text{for } x \in \mathfrak{g}.$$

Symbolically, $\bar{t}$ is a matrix of the form

$$\bar{t}=\begin{pmatrix} I_n & 0 \\ t & I_r \end{pmatrix},$$

where r is the dimension of $\mathfrak{g}$. Then G_1 is a vector group isomorphic to $\mathfrak{g}_1$.

We recall (see § 1) that the k-th prolongation $\mathfrak{g}_k$ of $\mathfrak{g}$ is the space of symmetric multilinear mappings

$$t\colon \underbrace{V\times\cdots\times V}_{k+1\text{-times}}\to V$$

such that, for each fixed $v_1, \dots, v_k\in V$, the linear transformation $v\in V\to t(v, v_1, \dots, v_k)\in V$ is in $\mathfrak{g}$. The *k-th prolongation* G_k of G is the group consisting of those linear transformations $\bar{t}$ of $V+\mathfrak{g}+\mathfrak{g}_1+\cdots+\mathfrak{g}_{k-1}$ induced by the elements t of $\mathfrak{g}_k$ as follows:

$$\bar{t}(v)=v+t(\cdot, \dots, \cdot, v) \qquad \text{for } v\in V,$$
$$\bar{t}(x)=x \qquad \text{for } x\in V+\mathfrak{g}+\mathfrak{g}_1+\cdots+\mathfrak{g}_{k-1}.$$

Symbolically, $\bar{t}$ is a matrix of the form

$$\bar{t}=\begin{pmatrix} I_n & 0 & 0 \\ 0 & I_N & 0 \\ t & 0 & I_r \end{pmatrix},$$

where $N=\dim(\mathfrak{g}+\mathfrak{g}_1+\cdots+\mathfrak{g}_{k-2})$ and $r=\dim \mathfrak{g}_{k-1}$. Then G_k is a vector group isomorphic to $\mathfrak{g}_k$.

It is easily seen that the first prolongation of G_{k-1} coincides with G_k. Thus G_k can be obtained from G by successive first prolongations.

Denote the dual space of V by V^*. Then $V\otimes\Lambda^2 V^*$ may be considered as the space of all skew-symmetric bilinear mappings from $V\times V$ into V, and similarly, $\mathfrak{g}\otimes V^*$ may be identified with the space of linear mappings from V into $\mathfrak{g}$. Define a linear mapping $\partial\colon \mathfrak{g}\otimes V^*\to V\otimes\Lambda^2 V^*$ by

$$(\partial f)(v_1, v_2)=-f(v_2)\,v_1+f(v_1)\,v_2 \qquad \text{for } f\in\mathfrak{g}\otimes V^*,\ v_1, v_2\in V.$$

It is clear that f is in the kernel of ∂ if and only if the mapping $(v_1, v_2)\in V\times V\to f(v_1)\,v_2\in V$ is in the first prolongation $\mathfrak{g}_1$.

We choose once and for all a linear subspace C of $V\otimes\Lambda^2 V^*$ such that

$$V\otimes\Lambda^2 V^*=\partial(\mathfrak{g}\otimes V^*)+C.$$

In general there is no natural way of choosing C.

Let P be a G-structure on an n-dimensional manifold M; it is a subbundle of $L(M)$ with structure group G. Let θ be the canonical form on P; it is a V-valued 1-form on P. An n-dimensional subspace H of the tangent space $T_u(P)$ at u is said to be horizontal if $\theta: H \to V$ is an isomorphism. The exterior derivative $d\theta$ evaluated at $u \in P$ is a skew-symmetric bilinear mapping $(d\theta)_u: T_u(P) \times T_u(P) \to V$. In view of the isomorphism $\theta: H \to V$, $d\theta$ restricted to $H \times H$ defines a skew-symmetric bilinear mapping $V \times V \to V$, i.e., an element $c(u, H) \in V \otimes \Lambda^2 V^*$, called the *torsion* corresponding to (u, H).

To discuss the dependence of $c(u, H)$ on H, we fix a pair (u, H) and choose a connection form ω on P such that H is horizontal with respect to ω, i.e., $\omega(X)=0$ for $X \in H$. The so-called first structure equation reads

$$d\theta = -\omega \wedge \theta + \Theta,$$

where Θ is the torsion form. Let $v_1, v_2 \in V$ and $X_1, X_2 \in H$ such that $\theta(X_i)=v_i$, $i=1,2$. Then $c(u, H)(v_1, v_2)=\Theta(X_1, X_2)$, thus justifying the name "torsion" for $c(u, H)$. Let H' be another horizontal subspace of $T_u(P)$ and $X_1', X_2' \in H'$ be vectors such that $\theta(X_i')=v_i$, $i=1,2$. Then, since $\Theta(X_1', X_2')=\Theta(X_1, X_2)$, we obtain

$$\begin{aligned} c(u, H')(v_1, v_2) - c(u, H)(v_1, v_2) &= d\theta(X_1', X_2') - d\theta(X_1, X_2) \\ &= -(\omega \wedge \theta)(X_1', X_2') + (\omega \wedge \theta)(X_1, X_2) \\ &= -\omega(X_1')\,\theta(X_2') + \omega(X_2')\,\theta(X_1'). \end{aligned}$$

If we define an element f of $\mathfrak{g} \otimes V^*$, i.e., a linear mapping from V into $\mathfrak{g}$ by

$$f(v) = \omega(X') \qquad v \in V,$$

where $X' \in H'$ is determined by $\theta(X')=v$, then

$$-\omega(X_1')\,\theta(X_2') + \omega(X_2')\,\theta(X_1') = -(\partial f)(v_1, v_2).$$

This shows that $c(u, H') - c(u, H)$ is an element of $\partial(\mathfrak{g} \otimes V^*)$.

Each horizontal subspace H of $T_u(P)$ determines a linear frame of the manifold P at u as follows. Since the structure group G acts on P, every element A of $\mathfrak{g}$ induces a vertical vector field A^* on P, called the fundamental vector field corresponding to A. Hence we have a linear mapping $\mathfrak{g} \to T_u(P)$ which sends A into A_u^*. On the other hand, we have a linear mapping $V \to H \subset T_u(P)$ which sends v into the vector $X \in H$ defined by $\theta(X)=v$. Adding these two linear mappings, we obtain a linear isomorphism $V + \mathfrak{g} \to T_u(P)$, which is by definition the linear frame of P determined by H. If we take a basis $e_1, \ldots, e_{n+r}$ in $V + \mathfrak{g}$ in such a way that $e_1, \ldots, e_n$ is the natural basis for $V = \mathbf{R}^n$ and $e_{n+1}, \ldots, e_{n+r}$ is a basis for $\mathfrak{g}$, then the image of this basis under the isomorphism $V + \mathfrak{g} \to T_u(P)$ is the linear frame determined by H (if one wants to define a linear frame at u as an ordered basis of $T_u(P)$ rather than a linear isomorphism from a fixed vector space $V + \mathfrak{g}$ onto $T_u(P)$).

We are now in a position to define the *first prolongation* P_1 of a G-structure P over M. The definition will depend on the fixed subspace C of $V \otimes \Lambda^2 V^*$ complementary to $\partial(\mathfrak{g} \otimes V^*)$. Let P_1 be the set of linear frames over P corresponding to horizontal subspaces $H \subset T_u(P)$ such that $u \in P$ and $c(u, H) \in C$. Then P_1 is a G_1-structure over P. In fact, if $a \in \mathrm{GL}(n+r; \mathbf{R})$ and $z \in P_1 \subset L(P)$, then $z \cdot a$ is defined by

$$z \cdot a\colon\ V+\mathfrak{g} \xrightarrow{a} V+\mathfrak{g} \xrightarrow{z} T_u(P),$$

and $z \cdot a$ is in P_1 if and only if $a \in G_1$. (To see this, let H and H' be the horizontal subspaces of $T_u(P)$ corresponding to z and $z \cdot a$, respectively. By definition, $c(u, H)$ is in C. We know that $c(u, H') - c(u, H)$ is in $\partial(\mathfrak{g} \otimes V^*)$. Since $z \cdot a$ is in P_1 if and only if $c(u, H')$ is in C, it follows that $z \cdot a$ is in P_1 if and only if $c(u, H') - c(u, H) \in C \cap \partial(\mathfrak{g} \otimes V^*) = 0$. This means that $z \cdot a$ is in P_1 if and only if the element f of $\mathfrak{g} \otimes V^*$ defined above is in the kernel of ∂. But the kernel of ∂ is precisely $\mathfrak{g}_1$. Our assertion is now immediate.)

The *k-th prolongation* P_k of P is defined inductively by $P_k = (P_{k-1})_1 =$ the first prolongation of P_{k-1}; it is a G_k-structure over P_{k-1}.

Let φ be an automorphism of a G-structure P over M; it is a transformation of M such that the induced bundle automorphism φ_* of $L(M)$ leaves P invariant. We denote the restriction of φ_* to P by φ_1. Then φ_1 is an automorphism of the G_1-structure P_1 over P. To see this, let H be a horizontal subspace of $T_u(P)$ such that $c(u, H) \in C$ so that the corresponding linear frame z of P at u is in P_1. From the fact that θ is invariant by φ_1, it follows that $c(\varphi_1(u), \varphi_{1*}(H)) = c(u, H) \in C$. Hence, the linear frame $\varphi_{1*}(z)$ corresponding to the horizontal subspace $\varphi_{1*}(H)$ is in P_1. This proves our assertion.

We can construct inductively a transformation φ_k of P_{k-1} which is an automorphism of the G_k-structure P_k over P_{k-1}.

Theorem 5.1. *Let P be a G-structure on an n-dimensional manifold M and $\mathfrak{A}$ the group of automorphisms of P. If the Lie algebra $\mathfrak{g} \subset \mathfrak{gl}(n; \mathbf{R})$ is of finite type of order k, then $\mathfrak{A}$ is a Lie transformation group of dimension $\leqq \dim(V+\mathfrak{g}+\mathfrak{g}_1+\cdots+\mathfrak{g}_{k-1})$.*

Proof. Since $\mathfrak{g}_k = 0$, G_k consists of the identity element only and the G_k-structure P_k over P_{k-1} is a $\{1\}$-structure, i.e., an absolute parallelism on P_{k-1}. Let $\mathfrak{A}_i$ denote the group of automorphisms of the G_i-structure P_i over P_{i-1}, i.e., transformations of P_{i-1} inducing automorphisms of P_i. By Theorem 3.2, $\mathfrak{A}_k$ is a Lie transformation group of dimension $\leqq \dim P_{k-1}$; for a fixed element z of P_{k-1}, the mapping $\psi \in \mathfrak{A}_k \to \psi(z) \in P_{k-1}$ imbeds $\mathfrak{A}_k$ as a closed submanifold of P_{k-1}. On the other hand, we can imbed $\mathfrak{A}$ into $\mathfrak{A}_k$ as a closed subset by mapping $\varphi \in \mathfrak{A}$ into $\varphi_k \in \mathfrak{A}_k$. Hence, $\mathfrak{A}$ is a Lie transformation group of dimension $\leqq \dim P_{k-1}$. Since $\dim P_{k-1} = \dim(V+\mathfrak{g}+\mathfrak{g}_1+\cdots+\mathfrak{g}_{k-1})$, the theorem is now proved. q.e.d.

Theorem 5.1 was first proved in special cases, e.g., for Riemannian conformal, projective structures and, more generally, for Cartan connections as an application of Theorem 3.2 (Kobayashi [1, 4]), for pseudogroup structures of finite type and hence for integrable G-structures of finite type (Ehresmann [4], Libermann [3]) and then in this general form by Ruh [1] and Sternberg [1].

6. Volume Elements and Symplectic Structures

We shall reconsider some of the examples discussed in § 2. Since we have established basic theorems on automorphism groups for elliptic G-structures and for G-structures of finite type in §§ 4 and 5, we shall be concerned with G-structures of non-elliptic infinite type in this section.

Let M be an n-dimensional manifold and consider the $\mathrm{GL}(n; \mathbf{R})$-structure on M, i.e., the bundle $L(M)$ of linear frames over M (see Example 2.1). The group of automorphisms of the $\mathrm{GL}(n; \mathbf{R})$-structure is nothing but the group of diffeomorphisms of M, which will be denoted by $\mathfrak{D}(M)$. Similarly, the Lie algebra of infinitesimal automorphisms of the $\mathrm{GL}(n; \mathbf{R})$-structure is the Lie algebra $\mathfrak{X}(M)$ of vector fields on M. When M is noncompact, practically nothing seems to be known about $\mathfrak{D}(M)$ and $\mathfrak{X}(M)$. For instance, a natural question would be whether $\mathfrak{D}(M)$ can be made into an infinite dimensional Lie group in a suitable sence. One of the difficulties seems to be lack of the corresponding Lie algebra. Since some vector fields are not complete, i.e., cannot be integrated globally, $\mathfrak{X}(M)$ is too large to be the Lie algebra of $\mathfrak{D}(M)$. On the other hand, the subset of $\mathfrak{X}(M)$ consisting of complete vector fields is not even a linear subspace of $\mathfrak{X}(M)$ (see Palais [1] for an example of two complete vector fields whose sum is not complete). Leslie [1] has shown that if M is compact, then $\mathfrak{D}(M)$ can be made into a Fréchet Lie group. Perhaps more useful is a strong ILH-Lie group structure introduced in $\mathfrak{D}(M)$ by Omori [1] in the case when M is compact (where ILH stands for "Inverse Limit of Hilbert"), see also Ebin-Marsden [1]. If one wishes to generalize Omori's results to the case of a noncompact manifold M, then the group to be considered is perhaps an appropriate completion $\overline{\mathfrak{D}}_c(M)$ of the subgroup $\mathfrak{D}_c(M)$ of $\mathfrak{D}(M)$ consisting of diffeomorphisms with compact support, i.e., transformations which act trivially outside compact sets, with Lie algebra $\mathfrak{X}_c(M)$ consisting of vector fields with compact support. Clearly, every element of $\mathfrak{X}_c(M)$ is a complete vector field and generates a 1-parameter subgroup of $\mathfrak{D}_c(M)$. The Lie algebra of $\overline{\mathfrak{D}}_c(M)$ would consist of vector fields decreasing rapidly at infinity in a suitable sense.

We shall now consider an $\mathrm{SL}(n; \mathbf{R})$-structure on M, i.e., a volume element μ on M (see Example 2.3). Let $\mathfrak{A}(M, \mu)$ (resp. $\mathfrak{a}(M, \mu)$) denote the group (resp. the Lie algebra) of transformations f (resp. infinitesimal transformations X) such that $f^*\mu=\mu$ (resp. $L_X\mu=0$). If M is compact, then $\mathfrak{A}(M,\mu)$ is a closed Fréchet Lie subgroup and hence a closed strong ILH-Lie subgroup of $\mathfrak{D}(M)$ with Lie algebra $\mathfrak{a}(M,\mu)$ (Omori [2], Ebin-Marsden [1]). Again it is possible that the correct group to be considered for a noncompact M is a completion $\overline{\mathfrak{A}}_c(M,\gamma)$ of the group $\mathfrak{A}_c(M,\mu)=\mathfrak{A}(M,\mu)\cap\mathfrak{D}_c(M)$ of μ-preserving transformations with compact support with the corresponding Lie algebra $\mathfrak{a}_c(M,\mu)=\mathfrak{a}(M,\mu)\cap \mathfrak{X}_c(M)$.

The following result is due to Boothby [3].

Theorem 6.1. *Given a volume element μ on a manifold M of dimension $n\geqq 2$, the group $\mathfrak{A}_c(M,\mu)$ of μ-preserving transformations with compact support (in fact, already the subgroup generated by $\mathfrak{a}_c(M,\mu)$) is k-fold transitive on M for every positive integer k.*

We recall that a group acting on M is said to be *k-fold transitive* if for arbitrary two sets of k distinct points $\{p_1,\dots,p_k\}$ and $\{q_1,\dots,q_k\}$ there is an element of the group which sends p_i into q_i for all $i=1,\dots,k$. Following Boothby we say that a group acting on M is *strongly locally transitive* if for each point $p\in M$ and each neighborhood U of p there are relatively compact neighborhoods V and W of p with $\overline{W}\subset V\subset\overline{V}\subset U$, and for each $q\in W$ there is an element of the group which leaves fixed every point outside $\overline{V}$ and sends p into q.

Proof. The following general lemma will be used again later.

Lemma 1. *If a group $\mathfrak{G}$ is strongly, locally transitive on M, it is k-fold transitive on M for every positive integer k.*

Proof of Lemma 1. Let $\{p_1,\dots,p_k\}$ and $\{q_1,\dots,q_k\}$ be two sets of k distinct points. For each i, let c_i be a curve from p_i to q_i chosen in such a way that $c_1,\dots,c_k$ are mutually disjoint. For each i, let N_i be a neighborhood of c_i chosen in such a way that $N_1,\dots,N_k$ are disjoint. It suffices to show that for each i there is an element g_i of $\mathfrak{G}$ which sends p_i into q_i and leaves every point outside N_i fixed. To obtain such an element g_i, for each point r on c_i we choose relatively compact neighborhoods V_r and W_r of r with $\overline{W}_r\subset V_r\subset\overline{V}_r\subset U=N_i$ in the manner described in the definition of "strongly locally transitive action". Then we choose points $p_i=r_0$, $r_1,\dots,r_m=q_i$ on c_i in such a way that $r_j\in W_{r_{j-1}}$ for $j=1,\dots,m$. Then we can send p_i successively to $r_1,\dots,r_m=q_i$ by elements of $\mathfrak{G}$ without disturbing the points outside N_i.

Lemma 2. *Let μ be a volume element on M. Then the group generated by $\mathfrak{a}_c(M,\mu)$ is strongly locally transitive on M.*

Proof of Lemma 2. Given a point $p \in M$ and a neighborhood U of p, let $x^1, \dots, x^n$ be a local coordinate system around p such that $\mu = dx^1 \wedge \dots \wedge dx^n$; such a coordinate system exists since every SL(n; **R**)-structure is integrable (see Example 2.3). Let V and W be neighborhoods of p with $\overline{W} \subset V \subset \overline{V} \subset U$ defined by

$$V:\ |x^i| < a \quad \text{and} \quad W:\ |x^i| < b, \qquad \text{where } 0 < b < a.$$

Let $q \in W$. Applying a linear change to the coordinate system, we may assume that q has coordinates $(c, 0, \dots, 0)$. Let f be a function with support in V such that f depends only on the variables x^1, x^2 and $f = x^2$ on W. Then define a vector field X by

$$X = \frac{\partial f}{\partial x^2} \frac{\partial}{\partial x^1} - \frac{\partial f}{\partial x^1} \frac{\partial}{\partial x^2}.$$

It is then easy to verify that X is a μ-preserving infinitesimal transformation with support in V and its orbit through the origin $(0, 0, \dots, 0)$ passes through the point q.

Now the theorem follows immediately from Lemmas 1 and 2. q.e.d.

We mention a theorem of Moser [1] which says that on a compact manifold all volume elements are essentially equivalent. This has been used in Omori's work mentioned above.

Theorem 6.2. *Let μ and ν be two volume elements on a compact manifold M. Then there is a transformation f of M such that $f^* \mu = \nu$ if and only if $\int_M \mu = \int_M \nu$.*

In connection with possible generalizations of Omori's results to the noncompact case, we should point out that Theorem 6.2 can be generalized to a noncompact M as follows. *Let μ and ν be two volume elements on M. Then there is a transformation $f \in \mathfrak{D}_c(M)$ with compact support such that $f^* \mu = \nu$ if and only if there is a compact subset K of M such that $\int_K \mu = \int_K \nu$ and $\mu = \nu$ outside K.*

We shall now consider a symplectic structure on a manifold M of dimension $n = 2m$, i.e., a closed 2-form ω of maximal rank (see Example 2.8). A transformation f (resp. an infinitesimal transformation X) of M is said to be *symplectic* if $f^* \omega = \omega$ (resp. $L_X \omega = 0$). We denote by $\mathfrak{A}(M, \omega)$ (resp. $\mathfrak{a}(M, \omega)$) the group (resp. the Lie algebra) of symplectic transformations (resp. infinitesimal symplectic transformations). If M is compact, then $\mathfrak{A}(M, \omega)$ is a closed Fréchet Lie subgroup and hence a closed strong ILH-Lie subgroup of $\mathfrak{D}(M)$ with Lie algebra $\mathfrak{a}(M, \omega)$ (Ebin-Marsden [1], Omori [2], Weinstein [2]). Again the question arises whether the group $\mathfrak{A}_c(M, \omega) = \mathfrak{A}(M, \omega) \cap \mathfrak{D}_c(M)$ of symplectic transformations with compact support is a strong ILH-Lie group with the corresponding Lie algebra $\mathfrak{a}_c(M,\omega)=\mathfrak{a}(M,\omega)\cap\mathfrak{X}_c(M)$ of infinitesimal symplectic transformations with compact support.

The following result is due to Libermann [3].

Theorem 6.3. *Let M be a symplectic manifold with closed 2-form ω of maximal rank. Under the linear isomorphism between the space $\mathfrak{X}(M)$ of vector fields and the space $\mathscr{A}^1(M)$ of 1-forms given by $X\in\mathfrak{X}(M)\to -\iota_X\,\omega\in\mathscr{A}^1(M)$, the space $\mathfrak{a}(M,\omega)$ of infinitesimal symplectic transformations is isomorphic to the space $\mathscr{C}^1(M)$ of closed 1-forms. Under the same isomorphism, the derived subalgebra $[\mathfrak{a}(M,\omega),\mathfrak{a}(M,\omega)]$ is mapped into the space $\mathscr{B}^1(M)$ of exact 1-forms.*

Proof. We apply the formula $L_X=d\circ\iota_X+\iota_X\circ d$ to ω (see for example Kobayashi-Nomizu [1; vol. 1, p. 35]). Since ω is closed, we obtain

$$L_X\,\omega=d\circ\iota_X\,\omega.$$

This shows that X is an infinitesimal symplectic transformation if and only if $\iota_X\,\omega$ is closed. To prove the second statement, we use the formula $\iota_{[X,Y]}=[L_X,\iota_Y]$ (see Kobayashi-Nomizu [1; vol. 1, p. 35]). Assume that X and Y are infinitesimal symplectic transformations. Since $L_X\,\omega=0$ and $d\circ c_Y\,\omega=L_Y\,\omega=0$, we obtain

$$\iota_{[X,Y]}\,\omega=L_X\circ\iota_Y\,\omega-\iota_Y\circ L_X\,\omega=d\circ\iota_X\circ\iota_Y\,\omega=2d\big(\omega(Y,X)\big).$$

This proves the second assertion. q.e.d.

This is probably an appropriate place to mention the classical Poisson bracket $\{f,g\}$. Let f and g be two functions on a symplectic manifold M. Let X_f and X_g be the vector fields corresponding to the exact 1-forms df and dg, respectively, under the duality defined by the symplectic form ω, i.e.,

$$df=-\iota_{X_f}\,\omega,\qquad dg=-\iota_{X_g}\,\omega.$$

We set

$$\{f,g\}=\iota_{X_g}\circ\iota_{X_f}\,\omega\,\big(=2\omega(X_f,X_g)\big).$$

It is easy to verify that the space $\mathscr{F}(M)$ of functions on M with Poisson bracket $\{\ ,\ \}$ is a Lie algebra. From the last formula in the proof of Theorem 6.3, it follows that $d\{f,g\}$ corresponds to $[X_f,X_g]$, i.e., $d\{f,g\}=-\iota_{[X_f,X_g]}\,\omega$. This fact implies that the mapping $f\to X_f$ defines a Lie algebra homomorphism from $\mathscr{F}(M)$ into $\mathfrak{a}(M,\omega)$. The kernel of this homomorphism consists of the constant functions. If we express ω in terms of an admissible coordinate system $x^1,\dots,x^{2m}$ as

$$\omega=dx^1\wedge dx^{m+1}+\dots+dx^m\wedge dx^{2m}$$

(see Example 2.8), then by a simple calculation we obtain the classical formula

$$\{f, g\} = \sum_{i=1}^{m} \frac{\partial f}{\partial x^i} \frac{\partial g}{\partial x^{m+i}} - \frac{\partial f}{\partial x^{m+i}} \frac{\partial g}{\partial x^i}.$$

The following result, proved first by Hatakeyama [1] in the compact case, is due to Boothby [3].

Theorem 6.4. *Given a symplectic manifold M with closed 2-form ω of maximal rank, the group $\mathfrak{A}_c(M, \omega)$ of symplectic transformations with compact support (in fact, already the subgroup generated by $\mathfrak{a}_c(M, \omega)$) is k-fold transitive on M for every positive integer k.*

Proof. In view of Lemma 1 in the proof of Theorem 6.1, it suffices to prove that the group generated by $\mathfrak{a}_c(M, \omega)$ is strongly locally transitive on M. Given a point $p \in M$ and a neighborhood U of p, let $x^1, \ldots, x^{2m}$ be a local coordinate system around p such that $\omega = dx^1 \wedge dx^{m+1} + \cdots + dx^m \wedge dx^{2m}$ (see Example 2.8). Let V and W be neighborhoods of p with $\overline{W} \subset V \subset \overline{V} \subset U$ defined by

$$V\colon\ |x^i| < a \quad \text{and} \quad W\colon\ |x^i| < b, \qquad \text{where } 0 < b < a.$$

Let $q \in W$. Applying a linear symplectic change to the coordinate system, we may assume that q has coordinates $(c, 0, \ldots, 0)$. Let f be a function with support in V such that $f = -x^{m+1}$ on W. Then the infinitesimal symplectic transformation X_f defined by $\iota_{X_f}\omega = df$ has support in V and coincides with $\partial/\partial x^1$ on W. Hence, the 1-parameter group generated by X_f maps the origin $p = (0, 0, \ldots, 0)$ into $q = (c, 0, \ldots, 0)$. q.e.d.

A result similar to Theorem 6.2 on the symplectic forms on a compact manifold is known but it is not as strong as Theorem 6.2, see Moser [1].

Given a symplectic structure on M with closed 2-form ω of maximal rank, a transformation f of M is said to be *conformal-symplectic* if $f^*\omega = \varphi \cdot \omega$, where φ is a function on M. Clearly, f is conformal-symplectic if and only if it is an automorphism of the $CSp(m; \mathbf{R})$-structure defined by ω (see Example 2.8). Since ω and $f^*\omega$ are closed, we obtain $d\varphi \wedge \omega = 0$. Making use of the expression $\omega = dx^1 \wedge dx^{m+1} + \cdots + dx^m \wedge dx^{2m}$, we can easily conclude that $d\varphi = 0$ if $m \geqq 2$. Hence (Libermann [3]),

Theorem 6.5. *If f is a conformal-symplectic transformation of a symplectic manifold M of dimension $2m \geqq 4$ with closed 2-form ω of maximal rank, then*

$$f^*\omega = c \cdot \omega,$$

where c is a nonzero constant. If M is compact, then $c = \pm 1$.

Proof. The last assertion follows from

$$\int_M f^*(\omega^m) = \pm \int_M \omega^m,$$

where the sign is positive or negative according as f is orientation-preserving or reversing. q.e.d.

An example of a symplectic manifold is a Kähler manifold with Kähler 2-form ω (whose complex structure and Riemannian structure are forgotten).

Another example is provided by the cotangent bundle $T^*(M)$ of any manifold M with the natural symplectic form ω defined as follows. Let γ be the 1-form on $T^*(M)$ defined by

$$\gamma(X) = \xi(\pi_* X) \qquad \text{for } X \in T_\xi(T^*(M)),$$

where $\pi\colon T^*(M) \to M$ is the natural projection so that $\pi_* X \in T_{\pi\xi}(M)$. Let $x^1, \ldots, x^n$ be a local coordinate system in M and $x^1, \ldots, x^n, p_1, \ldots, p_n$ be the induced local coordinate system in $T^*(M)$, i.e., $x^i(\xi) = x^i(\pi\,\xi)$ and $p_i(\xi) = \xi(\partial/\partial x^i)$ for $\xi \in T^*(M)$. Then

$$\gamma = \sum p_i\, dx^i.$$

Set

$$\omega = d\gamma = \sum dp_i \wedge dx^i.$$

For more results on automorphisms of (almost) symplectic and conformal-symplectic structures, see Lefevre [1–3], Libermann [2], Lichnerowicz [4]. On characterization of the symplectic structure on the cotangent bundle $T^*(M)$, see Nagano [11].

7. Contact Structures

Let M be a manifold of odd dimension $2m+1$. By a *contact form* on M we mean an open cover $\{U_i\}$ of M together with a system of 1-forms $\{\gamma_i\}$ such that

(1) each γ_i is a 1-form defined on U_i of maximal rank in the sense that $\gamma_i \wedge (d\gamma_i)^m \neq 0$ everywhere on U_i,

(2) we have $\gamma_i = f_{ij}\gamma_j$ on $U_i \cap U_j$, where f_{ij} is a function on $U_i \cap U_j$ (without zeros).

Two such forms $\{U_i, \gamma_i\}$ and $\{V_\lambda, \delta_\lambda\}$ are said to be equivalent if $\gamma_i = h_{i\lambda}\delta_\lambda$ on $U_i \cap V_\lambda$, where $h_{i\lambda}$ is a function on $U_i \cap V_\lambda$ (without zeros). An equivalence class of contact forms is called a *contact structure*. For simplicity's sake, we say "a contact structure $\{U_i, \gamma_i\}$" instead of "a contact structure represented by $\{U_i, \gamma_i\}$".

Given a contact structure $\{U_i, \gamma_i\}$, we define a vector subbundle of rank $2m$ of the tangent bundle $T(M)$ by setting

$$E_x = \{X \in T_x(M);\ \gamma_i(X) = 0\} \quad \text{if } x \in U_i.$$

From condition (1) it follows that $(d\gamma_i)^m \neq 0$ on the fibre E_x and hence $d\gamma_i$ defines a non-degenerate skew-symmetric bilinear form on E_x. This bilinear form on E_x is defined uniquely up to a nonzero constant multiple. It follows that the vector bundle E is orientable.

Let L be the quotient line bundle $T(M)/E$. Each contact form $\{U_i, \gamma_i\}$ gives rise to a globally defined 1-form with values in the line bundle L in the following manner. Since γ_i annihilates E_x, it can be considered as a linear functional on $T_x(M)/E_x$. Hence, the equation

$$\gamma_i(s_i) = 1$$

defines a cross section of L over U_i. Since $\gamma_i(s_i) = 1 = \gamma_j(s_j)$, we obtain

$$s_j = f_{ij}\, s_i \quad \text{on} \quad U_i \cap U_j.$$

It follows that the form $\tilde{\gamma}$ defined by

$$\tilde{\gamma} = \gamma_i\, s_i$$

is a globally defined 1-form with values in L. It is easy to verify that an equivalent contact form gives rise to the same L-valued 1-form $\tilde{\gamma}$. Thus $\tilde{\gamma}$ depends only on the contact structure defined by $\{U_i, \gamma_i\}$.

Given a contact structure $\{U_i, \gamma_i\}$ on M, a transformation f of M is called a *contact transformation* if $\{f^{-1} U_i, f^* \gamma_i\}$ and $\{U_i, \gamma_i\}$ are equivalent. Consequently, an *infinitesimal contact transformation* X is defined by the condition

$$L_X \gamma_i = g_i \cdot \gamma_i \qquad (\text{where } g_i \text{ is a function on } U_i).$$

More geometrically, f is a contact transformation if and only if the induced bundle automorphism f_*: $T(M) \to T(M)$ sends the subbundle E into itself, that is, if and only if f is an automorphism of the $GL(n-1, 1; \mathbf{R})$-structure on M defined by the subbundle E (see Example 2.9).

Writing γ for a contact structure $\{U_i, \gamma_i\}$ for simplicity's sake, we denote the group of contact transformations by $\mathfrak{A}(M, \gamma)$ and the Lie algebra of infinitesimal contact transformations by $\mathfrak{a}(M, \gamma)$. If M is compact, $\mathfrak{A}(M, \gamma)$ is an ILH-Lie group, Omori [3]. As in the preceding section, we denote the subgroup (resp. subalgebra) of $\mathfrak{A}(M, \gamma)$ (resp. $\mathfrak{a}(M, \gamma)$) consisting of elements with compact support by $\mathfrak{A}_c(M, \gamma)$ (resp. $\mathfrak{a}_c(M, \gamma)$).

Theorem 7.1. *Given a contact structure* $\{U_i, \gamma_i\}$ *on* M, *let* $\tilde{\gamma}$ *be the corresponding* 1-*form on* M *with values in the real line bundle* $L = T(M)/E$, *where* E *is the subbundle of* $T(M)$ *defined by* $\gamma_i = 0$. *Then the mapping* $X \in \mathfrak{a}(M, \gamma) \to \tilde{\gamma}(X) \in H^0(M; L)$ *gives a linear isomorphism from the space* $\mathfrak{a}(M, \gamma)$ *of infinitesimal contact transformations onto the space* $H^0(M; L)$ *of cross sections of the line bundle* L *over* M.

Proof. To prove that the mapping is injective, let $X \in \mathfrak{a}(M, \gamma)$ and assume $\tilde{\gamma}(X) = 0$, i.e., $\gamma_i(X) = 0$. Then

$$\iota_X \circ d\gamma_i = \iota_X \circ d\gamma_i + d \circ \iota_X \gamma_i = L_X \gamma_i = g_i \cdot \gamma_i.$$

Let $Y \in E_x \subset T_x(M)$ at $x \in U_i$. Then

$$\iota_Y \circ \iota_X \circ d\gamma_i = g_i \cdot (\iota_Y \gamma_i) = 0.$$

Since X is in E by our assumption $\gamma_i(X) = 0$ and since $d\gamma_i$ defines a non-degenerate bilinear form on E_x, we may conclude that $X = 0$. To prove that the mapping is surjective, let s be a cross section of L. As before, let s_i be the cross section of L over U_i defined by $\gamma_i(s_i) = 1$ and write

$$s = h_i s_i,$$

where h_i is a function on U_i. Let S_i be the vector field on U_i defined by

$$\gamma_i(S_i) = 1 \quad \text{and} \quad \iota_{S_i} \circ d\gamma_i = 0.$$

The projection $T(M) \to L = T(M)/E$ maps S_i into s_i. The vector field X we are looking for must be of the form

$$X = h_i S_i + Y_i,$$

where Y_i is a vector field on U_i contained in E. Since X is an infinitesimal contact transformation if and only if

$$\iota_X \circ d\gamma_i + d \circ \iota_X \gamma_i = g_i \cdot \gamma_i$$

and since a 1-form is a multiple of γ_i if and only if it annihilates E, a necessary and sufficient condition for X to be an infinitesimal contact transformation is that

$$\iota_Z \circ \iota_X \circ d\gamma_i + \iota_Z \circ d \circ \iota_X \gamma_i = 0 \qquad \text{for all vectors } Z \in E_x, \ x \in U_i.$$

But this is equivalent to

$$\iota_Z \circ \iota_{Y_i} \circ d\gamma_i + Z(h_i) = 0 \qquad \text{for } Z \in E_x, \ x \in U_i.$$

Since $d\gamma_i$ defines a non-degenerate bilinear form on E_x, this equation determines a unique vector $Y_i \in E_x$ at each $x \in U_i$. q.e.d.

We shall now see, under the isomorphism $\mathfrak{a}(M,\gamma)\to H^0(M;L)$, how the Lie bracket looks like in $H^0(M;L)$. Let X and Y be two infinitesimal contact transformations and let s and t be the corresponding sections of the line bundle L, i.e.,

$$s=\gamma(X)=\gamma_i(X)\,s_i \quad \text{and} \quad t=\gamma(Y)=\gamma_i(Y)\,s_i.$$

Setting

$$f=\gamma_i(X) \quad \text{and} \quad g=\gamma_i(Y)$$

and defining $[f,g]$ by

$$[f,g]=\gamma_i([X,Y]),$$

we want to express $[f,g]$ in terms of f and g. Since X is an intinitesimal contact transformation, we have $L_X\gamma_i=h_i\cdot\gamma_i$, where h_i is a function defined on U_i. We shall first express h_i in terms of f. Let S_i be the vector field on U_i constructed in the proof of Theorem 7.1. Then $X=fS_i+X'$, where X' is a vector field on U_i with values in E. We have

$$h_i\cdot\gamma_i=L_X\gamma_i=d\circ\iota_X\gamma_i+\iota_X\circ d\gamma_i=df+\iota_{X'}\circ d\gamma_i.$$

Applying ι_{S_i} to the both ends of the equalities, we obtain $h_i=S_i(f)$. Thus,

$$L_X\gamma_i=df+\iota_X\circ d\gamma_i=S_i(f)\cdot\gamma_i.$$

This formula allows us to recover X from f as follows. Taking U_i sufficiently small we may assume that γ_i is expressed as

$$\gamma_i=x^1\,dx^{m+1}+x^2\,dx^{m+2}+\dots+x^m\,dx^{2m}+dx^{2m+1}$$

in terms of an admissible local coordinate system $x^1,\dots,x^{2m+1}$ (see Appendix 1, Theorem 1). Then

$$d\gamma_i=dx^1\wedge dx^{m+1}+\dots+dx^m\wedge dx^{2m}$$

and

$$S_i=\partial/\partial x^{2m+1}.$$

By a direct calculation we obtain

$$X=\sum_{k=1}^{m}\left\{(x^k f_{2m+1}-f_{m+k})\frac{\partial}{\partial x^k}+f_k\frac{\partial}{\partial x^{m+k}}\right\}+\left(f-\sum_{k=1}^{m}x^k f_k\right)\frac{\partial}{\partial x^{2m+1}}$$

where

$$f_h=\partial f/\partial x^h \quad \text{for } h=1,\dots,2m+1.$$

From this expression we obtain

$$[f,g]=\sum_{k=1}^{m}\{f_k(g_{m+k}-x^k g_{2m+1})-g_k(f_{m+k}-x^k f_{2m+1})\}$$
$$+fg_{2m+1}-gf_{2m+1}.$$

The following result, proved first by Hatakeyama [1] in the compact case, is due to Boothby [3].

Theorem 7.2. *Given a contact structure $\{U_i, \gamma_i\}$ on a manifold M, the group $\mathfrak{A}_c(M,\gamma)$ of contact transformations with contact support (in fact, already the subgroup generated by $\mathfrak{a}_c(M,\gamma)$) is k-fold transitive on M for every positive integer k.*

Proof. In view of Lemma 1 in the proof of Theorem 6.1, it suffices to prove that the group generated by $\mathfrak{a}_c(M,\gamma)$ is strongly locally transitive. Given a point $p\in M$ and a neighborhood U of p, let $x^1, \dots, x^{2m+1}$ be a local coordinate system around p such that $\gamma_i = x^1\,dx^{m+1} + \dots + x^m\,dx^{2m} + dx^{2m+1}$ (assuming that $p\in U_i$). Let V be a small neighborhood of p with $\bar{V}\subset U$ such that the above expression of γ_i in terms of the local coordinate system is valid. We shall construct a vector subspace $\mathfrak{b}$ of $\mathfrak{a}_c(M,\gamma)$ of dimension $2m+1$ ($=\dim M$) consisting of vector fields with support in V such that the mapping

$$X\in\mathfrak{b}\to(\exp X)\,p\in M$$

gives a diffeomorphism of a neighborhood of 0 in $\mathfrak{b}$ onto a neighborhood of p in M. Let B be the $(2m+1)$-dimensional space of functions on V defined by

$$B=\left\{f=a_{2m+1}+\sum_{k=1}^{m}(a_{m+k}\,x^k-a_k\,x^{m+k});\ (a_1,\dots,a_{2m+1})\in\mathbf{R}^{2m+1}\right\}.$$

Let ρ be a function which is equal to 1 in a neighborhood of p and has support contained in V. Set $B_V=\{\rho f;\ f\in B\}$. Then B_V is a $(2m+1)$-dimensional space of functions on M with support in V. Using the notation in the proof of Theorem 7.1 and multiplying each element of B_V by s_i, we consider B_V as a subspace of the space $H^0(M;L)$ of sections of the line bundle L. Let $\mathfrak{b}$ be the subspace of $\mathfrak{a}(M,\gamma)$ corresponding to B_V under the isomorphism $\mathfrak{a}(M,\gamma)=H^0(M;L)$ established in Theorem 7.1. Using the explicit formula above which reproduces a vector field X from f, we see that a function

$$f=a_{2m+1}+\sum_{k=1}^{m}(a_{m+k}\,x^k-a_k\,x^{m+k})$$

gives rise to

$$\begin{aligned}X&=\sum_{k=1}^{m}\left\{(x^k f_{2m+1}-f_{m+k})\frac{\partial}{\partial x^k}+f_k\frac{\partial}{\partial x^{m+k}}\right\}+\left(f-\sum_{k=1}^{m}x^k f_k\right)\frac{\partial}{\partial x^{2m+1}}\\&=\sum_{k=1}^{m}\left(a_k\frac{\partial}{\partial x^k}+a_{m+k}\frac{\partial}{\partial x^{m+k}}\right)+\left(a_{2m+1}-\sum_{k=1}^{m}a_k\,x^{m+k}\right)\frac{\partial}{\partial x^{2m+1}}.\end{aligned}$$

Consider the differential of the mapping $\rho f \in B_V$ $(=\mathfrak{b}) \to (\exp X) p \in M$ at the origin 0. It is a linear mapping of $B_V = T_0(B_V)$ into $T_p(M)$ which sends $\rho f \in B_V$ into $X_p \in T_p(M)$. The formula above for X shows that $X_p = \sum_{j=1}^{2m+1} a_j \left(\frac{\partial}{\partial x^j}\right)_p$. Hence, the differential of $B_V \to M$ at the origin is non-degenerate. The theorem now follows from the inverse function theorem. q.e.d.

We remark here that the use of the subbundle E and the quotient line bundle L of $T(M)$ gives a simple proof of the following well known result, (cf. Gray [1]).

Theorem 7.3. *A contact structure $\{U_i, \gamma_i\}$ on M can be represented by a globally defined 1-form if and only if M is orientable.*

Proof. Since E is an orientable vector bundle, M is orientable (i.e., the tangent bundle $T(M)$ is orientable) if and only if the quotient bundle L is orientable. Since $L$is a real line bundle, it is orientable if and only if it has a cross section without zeros. If s is such a section, then $s_i = h_i s$, where h_i is a function on U_i and s_i is the section of L over U_i characterized by $\gamma_i(s_i) = 1$ (see the proof of Theorem 7.1). Then the 1-form $\gamma = h_i \gamma_i$ is well defined on M and satisfies the equation $\gamma(s) = 1$. Conversely, if the contact structure $\{U_i, \gamma_i\}$ can be represented by a globally defined 1-form γ, then $\gamma = h_i \gamma_i$ with a suitable function h_i on U_i. The section s defined by $s_i = h_i s$ is globally defined on M and satisfies $\gamma(s) = 1$. q.e.d.

An example of a contact manifold is an odd-dimensional sphere S^{2m+1}. In $\mathbf{C}^{m+1}$ with natural coordinate system $z^1, \ldots, z^{m+1}$, where $z^k = x^k + i y^k$, set $\gamma = \sum x^k dy^k$. Let S^{2m+1} be the unit sphere centered at the origin in $\mathbf{C}^{2m+1}$. Then γ induces a contact form on S^{2m+1}.

Another example is the cotangent sphere bundle over any manifold M. Let M be an $(m+1)$-dimensional manifold and $T^*(M)$ be the cotangent bundle. Let γ be the 1-form on $T^*(M)$ constructed at the end of § 6. Choose any Riemannian metric on M and let $S^*(M)$ be the unit sphere bundle consisting of covectors of length 1. Then γ induces a contact form on $S^*(M)$.

For more information on contact structures, see Boothby-Wang [1], Gray [1], Libermann [2], Sasaki [1], Takizawa [1], Lichnerowicz [4].

8. Pseudogroup Structures, G-Structures and Filtered Lie Algebras

As we have seen in § 2, the concept of G-structure unifies a large number of interesting geometric structures. We shall now consider another unifying concept, namely, that of pseudogroup structure.

Let E be an n-dimensional manifold which will be taken as a *model space*. It is usually a Euclidean space. A *pseudogroup of transformations* on E is a set Γ of local diffeomorphisms satisfying the following conditions:

(1) Each $f \in \Gamma$ is a diffeomorphism of an open set (called the domain of f) of E onto an open set (called the range of f) of E.

(2) Let $U = \bigcup_i U_i$, where each U_i is an open set of E. A diffeomorphism f of U onto an open set of E belongs to Γ if and only if the restriction of f to each U_i is in Γ.

(3) For every open set U of E, the identity transformation of U is in Γ.

(4) If f is in Γ, then f^{-1} is in Γ.

(5) If $f \in \Gamma$ is a diffeomorphism of U onto V and $f' \in \Gamma$ is a diffeomorphism of U' onto V' and if $V \cap U'$ is nonempty, then the diffeomorphism $f' \circ f$ of $f^{-1}(V \cap U')$ onto $f'(V \cap U')$ is in Γ.

A pseudogroup Γ of transformations of E is said to be *transitive* if for every pair of points p and q of E, there exists an element f of Γ such that $f(p) = q$.

Fix a transitive pseudogroup Γ of transformations of E. An *atlas* of a topological space M compatible with Γ (a *Γ-atlas*, for short) is a family of pairs (U_i, φ_i), called *charts*, such that

(a) Each U_i is an open set of M and $\bigcup_i U_i = M$.

(b) Each φ_i is a homeomorphism of U_i onto an open set of E.

(c) Whenever $U_i \cap U_j$ is nonempty, the mapping $\varphi_j \circ \varphi_i^{-1}$ of $\varphi_i(U_i \cap U_j)$ onto $\varphi_j(U_i \cap U_j)$ is an element of Γ.

A Γ-atlas of M is said to be *maximal* (or *complete*) if it is not contained in any other Γ-atlas of M. Every Γ-atlas is contained in a unique maximal Γ-atlas. A *Γ-structure* on M is a maximal Γ-atlas of M. It is customary to assume that M is a Hausdorff space. A *Γ-manifold* is a Hausdorff space M with a fixed maximal Γ-structure. Every Γ-atlas of M, enlarged to a unique maximal Γ-atlas, defines a Γ-structure on M.

We shall give a few examples.

Example 8.1. Let $E = \mathbf{R}^n$ and Γ be the set of all local diffeomorphisms of E. Then a Γ-manifold M is a usual (differentiable) manifold and a Γ-structure is a differentiable structure.

Example 8.2. Let $E = \mathbf{R}^n$ with natural coordinate system $x^1, \ldots, x^n$ and set $\omega = dx^1 \wedge \cdots \wedge dx^n$. Let Γ be the set of all local diffeomorphisms f of E such that $f^* \omega = c_f \cdot \omega$, where c_f is a constant (which depends on f). Given a Γ-atlas $\{(U_i, \varphi_i)\}$ of M, set $\omega_i = \varphi_i^* \omega$. Then each ω_i is an n-form on U_i without zeros and $\omega_i = c_{ij} \omega_j$ on $U_i \cap U_j$, where c_{ij} is a (nonzero) constant. Conversely, given a family of pairs (U_i, ω_i) such that U_i is an open cover of M and ω_i is an n-form on U_i without zeros satisfying $\omega_i = c_{ij}\,\omega_j$ on $U_i \cap U_j$, we can recover a Γ-atlas of M and a unique Γ-structure on M.

Example 8.3. Let E and ω be as in Example 8.2 and Γ be the set of all local diffeomorphisms f of E such that $f^*\omega = \omega$. Then the Γ-structures on M are in one-to-one correspondence with the volume elements of M in a natural manner. This structure was discussed in § 6.

Example 8.4. Let $E = \mathbf{R}^{2m}$ with natural coordinate system $x^1, \ldots, x^{2m}$ and set $\omega = dx^1 \wedge dx^{m+1} + \cdots + dx^m \wedge dx^{2m}$. Let Γ be the set of all local diffeomorphisms f of E such that $f^*\omega = c_f \cdot \omega$, where c_f is a constant. Then the Γ-structures are in a natural one-to-one correspondence with the conformal symplectic structures (see Example 2.8 and also § 6), provided $m \geqq 2$. We recall that if f is a local diffeomorphism such that $f^*\omega = \varphi_f \cdot \omega$ for some function φ_f, then φ_f is necessarily a constant.

Example 8.5. Let E and ω be as in Example 8.4 and Γ be the set of all local diffeomorphisms f of E such that $f^*\omega = \omega$. Then the Γ-structures are in a natural one-to-one correspondence with the symplectic structures discussed in Example 2.8 and in § 6.

Example 8.6. Let $E = \mathbf{R}^{2m+1}$ with natural coordinate system $x^1, \ldots, x^{2m+1}$ and set $\gamma = x^1\, dx^{m+1} + \cdots + x^m\, dx^{2m} + dx^{2m+1}$. Let Γ be the set of all local diffeomorphisms f of E such that $f^*\gamma = \varphi_f \cdot \gamma$, where φ_f is a function. Then the Γ-structures are in a natural one-to-one correspondence with the contact structures discussed in § 7.

Example 8.7. Let $E = \mathbf{R}^n$ and G be a Lie subgroup of $\mathrm{GL}(n; \mathbf{R})$. Let Γ be the set of all local diffeomorphisms f of E such that at each point of the domain of f the Jacobian matrix J_f belongs to G. Then the Γ-structures are in a natural one-to-one correspondence with the integrable G-structures.

Example 8.8. Let $E = \mathbf{R}^n$ and $G \subset \mathrm{GL}(n; \mathbf{R})$ as in Example 8.7. Let Γ be the set of all local diffeomorphisms f of E such that the Jacobian matrix J_f of f is constant and belongs to G. Then a Γ-structure is called a *flat G-structure*. If $G = \mathrm{GL}(n; \mathbf{R})$, a Γ-structure is known as an *affine structure*.

Example 8.9. Let E be a manifold on which a Lie group L is acting transitively and Γ be the set of all local diffeomorphisms f of E which can be obtained by localizing the elements of L. This example generalizes Example 8.8; taking $E = \mathbf{R}^n$ and $L = \mathbf{R}^n \cdot G$ (= the subgroup of the group of affine transformations generated by the translations $\mathbf{R}^n$ and a linear group G), we recover Example 8.8. If $E = S^n$ and L is the group of Möbius transformations (i.e., conformal transformations) of S^n, then a Γ-structure is called a *flat conformal structure.* If $E=P_n(\mathbf{R})$ and $L=\mathrm{PGL}(n;\mathbf{R})$ (=the projective general linear group), then a Γ-structure is called a *flat projective structure.*

In order to relate pseudogroup structures to G-structures, we shall consider transitive Lie pseudogroups. We shall not be concerned here with intransitive Lie pseudogroups; for various definitions of general Lie pseudogroups, see Ehresmann [3], Kuranishi [1], Libermann [3], Singer-Sternberg [1], Rodrigues [2].

Following Ehresmann we construct the bundle of r-frames over M, (cf. Kobayashi [8]). We fix a point of a model space E as the origin and denote it by 0. If $E=\mathbf{R}^n$, we take the usual origin 0 of $\mathbf{R}^n$. If V and V' are two neighborhoods of the origin 0 in E and if U and U' are two neighborhoods of a point $x\in M$, two diffeomorphisms $f\colon V\to U$ and $f'\colon V'\to U'$ are said to define the same *r-frame* at x if $x=f(0)=f'(0)$ and if f and f' have the same partial derivatives up to order r at 0 (in terms of local coordinate systems around 0 and x). The r-frame given by f is usually denoted by $j_0^r(f)$. The set of r-frames of M, denoted by $P^r(M)$, is a principal bundle over M with natural projection π, $\pi(j_0^r(f))=f(0)$, and with structure group $G^r(n)$ which will be now described. Let $G^r(n)$ be the set of r-frames $j_0^r(g)$ at $0\in E$; it forms a group with multiplication defined by

$$j_0^r(g)\circ j_0^r(g')=j_0^r(g\circ g').$$

The group $G^r(n)$ acts on $P^r(M)$ on the right by

$$j_0^r(f)\circ j_0^r(g)=j_0^r(f\circ g) \quad \text{for } j_0^r(f)\in P^r(M),\ j_0^r(g)\in G^r(n).$$

Clearly, $P^1(M)$ is the bundle of linear frames over M with group $G^1(n)=\mathrm{GL}(n;\mathbf{R})$.

Let Γ be a transitive pseudogroup on E and fix a Γ-structure on M. Let $P^r(M,\Gamma)$ be the subset of $P^r(M)$ consisting of r-frames $j_0^r(f)$ such that $f^{-1}\colon U\to V$ is a chart of the maximal Γ-atlas. Let $G^r(\Gamma)$ be the subgroup of $G^r(n)$ consisting of r-frames $j_0^r(g)$ at $0\in E$ such that $g\in\Gamma$. If $P^r(M,\Gamma)$ is a submanifold of $P^r(M)$, then $P^r(M,\Gamma)$ is a subbundle of $P^r(M)$ with group $G^r(\Gamma)$. Since E carries a natural Γ-structure itself, we can apply the construction of $P^r(M,\Gamma)$ to E to obtain $P^r(E,\Gamma)$. Since M and E are locally isomorphic as Γ-manifolds, $P^r(M,\Gamma)$ is a submanifold of $P^r(M)$ if (and only if) $P^r(E,\Gamma)$ is a submanifold of $P^r(E)$.

We say that a transitive pseudogroup Γ on E is a *Lie pseudogroup* if $P^r(E,\Gamma)$ is a submanifold (and hence a subbundle) of $P^r(E)$ for every positive integer r and if there is a positive integer s with the property that a local diffeomorphism h of E is in Γ if the induced local automorphism h_* of $P^s(E)$ leaves $P^s(E,\Gamma)$ invariant. (Roughly speaking, $P^s(E,\Gamma)$ is a system of partial differential equations and the condition says that if h sends solutions into solutions, then h must be in Γ.) The smallest integer s satisfying the above condition is called the *degree* of Γ.

Let Γ be a transitive Lie pseudogroup of degree s on E and fix a Γ-structure on M. Then we have a subbundle $P^s(E,\Gamma)$ of $P^s(E)$ and a subbundle $P^s(M,\Gamma)$ of $P^s(M)$. From $P^s(E,\Gamma)$ we can recover the pseudogroup Γ by taking all local diffeomorphisms of E which leave $P^s(E,\Gamma)$ invariant, i.e., all local automorphisms of $P^s(E,\Gamma)$. From $P^s(M,\Gamma)$ we can reconstruct the Γ-structure of M by taking all local diffeomorphisms of M into E which maps $P^s(M,\Gamma)$ into $P^s(E,\Gamma)$, i.e., all local isomorphisms of $P^s(M,\Gamma)$ into $P^s(E,\Gamma)$.

In order to unify the concept of G-structure and that of transitive Lie pseudogroup, we introduce the concept of G-structure of higher degree. Let G be a Lie subgroup of $G^s(n)$. Then a subbundle P of $P^s(M)$ with structure group G is called a *G-structure of degree s* on M. A G-structure in the sense of earlier sections is a G-structure of degree 1. Given $G\subset G^s(n)$, let $\mathbf{R}^n\times G$ denote the natural (flat) G-structure of degree s on $\mathbf{R}^n$. A diffeomorphism f of an open set U of M onto an open set of $\mathbf{R}^n$ is called an *admissible local coordinate system* if f induces an isomorphism of $P|_U$ onto $(\mathbf{R}^n\times G)|_{f(U)}$. If every point of M has a neighborhood with admissible local coordinate system, then the G-structure P is said to be *integrable*. This generalizes the concept of integrability introduced in § 1.

We shall now reexamine Examples 8.1 through 8.9. The pseudogroups Γ in these examples are all transitive Lie pseudogroups. Those in Examples 8.1, 8.3, 8.4 (for $m\geqq 2$), 8.5, 8.6, 8.7 are of degree 1 and, in each of these cases, $G^1(\Gamma)$ is $\mathrm{GL}(n;\mathbf{R})$, $\mathrm{SL}(n;\mathbf{R})$, $\mathrm{CSp}(m;\mathbf{R})$, $\mathrm{Sp}(m;\mathbf{R})$, $\mathrm{GL}(n-1,1;\mathbf{R})$ and G, respectively. The pseudogroups Γ in Examples 8.2 and 8.8 are of degree 2. In case of Example 8.9, the degree s of Γ is the smallest integer with the following property: if g and g' are transformations of E given by elements of L, $j_0^{s-1}(g)=j_0^{s-1}(g')$ implies $g=g'$. The $G^s(\Gamma)$-structure $P^s(M,\Gamma)$ is integrable in Examples 8.1, 8.2, 8.3, 8.4, 8.5, 8.7, 8.8 while it is never integrable in Example 8.6. In case of Example 8.9, $P^s(M,\Gamma)$ is sometimes integrable, e.g., when L is the group of Möbius transformations of S^n or when L is $\mathrm{PGL}(n;\mathbf{R})$ acting on $P_n(\mathbf{R})$.

To each transitive Lie pseudogroup Γ we shall associate a transitive filtered Lie algebra. In general, we define a *filtered Lie algebra* to be a Lie algebra $\mathfrak{l}$ (possibly of infinite dimension) with decreasing sequence of subalgebras $\mathfrak{l}=\mathfrak{l}_{-1}\supset\mathfrak{l}_0\supset\mathfrak{l}_1\supset\cdots$ such that

(a) $\quad [\mathfrak{l}_p,\mathfrak{l}_q]\subset\mathfrak{l}_{p+q}$,

(b) $\quad \dim \mathfrak{l}_p/\mathfrak{l}_{p+1}<\infty$,

(c) $\quad \bigcap_p \mathfrak{l}_p=0.$

A filtered Lie algebra $\mathfrak{l}$ is said to be *transitive* if

(d) $\quad \mathfrak{l}_p=\{X\in\mathfrak{l}_{p-1};\ [X,\mathfrak{l}]\subset\mathfrak{l}_{p-1}\}\quad$ for $p\geqq 1$.

As we have seen earlier, a *graded Lie algebra* is a Lie algebra $\mathfrak{g} = \sum_{p=-1}^{\infty} \mathfrak{g}_p$ such that

(a′) $$[\mathfrak{g}_p, \mathfrak{g}_q] \subset \mathfrak{g}_{p+q},$$

(b′) $$\dim \mathfrak{g}_p < \infty.$$

A graded Lie algebra $\mathfrak{g}$ is said to be *transitive* if

(d′) $[X, \mathfrak{g}_{-1}] \neq 0$ for every nonzero $X \in \mathfrak{g}_p$, $p \geqq 0$.

Every (transitive) graded Lie algebra $\mathfrak{g}$ may be considered as a (transitive) filtered Lie algebra in a natural manner, i.e., $\mathfrak{l}_p = \mathfrak{g}_p + \mathfrak{g}_{p+1} + \cdots$. To each (transitive) filtered Lie algebra $\mathfrak{l}$ we can associate a graded Lie algebra $\mathfrak{g}(\mathfrak{l}) = \sum \mathfrak{g}_p(\mathfrak{l})$ by setting $\mathfrak{g}_p(\mathfrak{l}) = \mathfrak{l}_p/\mathfrak{l}_{p+1}$ and defining the bracket operation in a natural manner. Two non-isomorphic filtered Lie algebras may give rise to the same graded Lie algebra.

Given a transitive Lie pseudogroup Γ acting on E, let $\mathfrak{a}$ be the Lie algebra of germs of vector fields X at $0 \in E$ such that $\exp(tX)$ is in Γ for small values of t, $|t| < \delta$. Let $\mathscr{F}$ denote the algebra of germs of functions defined around $0 \in E$ and $\mathscr{I}$ be the maximal ideal of $\mathscr{F}$, i.e., the ideal consisting of germs of functions vanishing at 0. We define a filtration $\mathfrak{a} = \mathfrak{a}_{-1} \supset \mathfrak{a}_0 \supset \mathfrak{a}_1 \supset \ldots$ by

$$\mathfrak{a}_p = \{X \in \mathfrak{a};\ X(\mathscr{F}) \subset \mathscr{I}^{p+1}\}.$$

In other words, X is in $\mathfrak{a}_p$ if and only if the components of X, expanded into Taylor series in terms of a local coordinate system around $0 \in E$, have no terms of degree less than p. Then with this filtration the Lie algebra $\mathfrak{a}$ satisfies conditions (a), (b) and (d). Set $\mathfrak{a}_\infty = \bigcap_p \mathfrak{a}_p$ and define $\mathfrak{l} = \mathfrak{a}/\mathfrak{a}_\infty$ and $\mathfrak{l}_p = \mathfrak{a}_p/\mathfrak{a}_\infty$. Then the Lie algebra $\mathfrak{l}$ with filtration $\mathfrak{l} = \mathfrak{l}_{-1} \supset \mathfrak{l}_0 \supset \mathfrak{l}_1 \ldots$ is a transitive filtered Lie algebra. We note that $\mathfrak{a}_\infty$ consists of germs of vector fields X such that the Taylor series of the components of X (when expanded in terms of a local coordinate system around $0 \in E$) are trivial.

This filtered Lie algebra $\mathfrak{l}$ is useful in studying the automorphisms of a Γ-structure. But we shall not go into this question here.

For filtered and graded Lie algebras, see E. Cartan [5–7], Guillemin [2], Guillemin-Sternberg [1], Kac [1], Kobayashi-Nagano [3, 4], Ochiai [1], Singer-Sternberg [1], Shnider [1], Tanaka [5–7], Weisfeiler [1], Morimoto-Tanaka [1].

II. Isometries of Riemannian Manifolds

1. The Group of Isometries of a Riemannian Manifold

The earliest and very general result on the group of isometries is perhaps the following theorem of van Danzig and van der Waerden [1] (see also Kobayashi-Nomizu [1, vol. 1; pp. 46–50] for a proof).

Theorem 1.1. *Let M be a connected, locally compact metric space and* $\mathfrak{I}(M)$ *the group of isometries of M. For each point x of M, let* $\mathfrak{I}_x(M)$ *denote the isotropy subgroup of* $\mathfrak{I}(M)$ *at x. Then* $\mathfrak{I}(M)$ *is locally compact with respect to the compact-open topology and* $\mathfrak{I}_x(M)$ *is compact for every x. If M is compact, then* $\mathfrak{I}(M)$ *is compact.*

Eleven years later, in 1939, the following result was published by Myers and Steenrod [1].

Theorem 1.2. *The group* $\mathfrak{I}(M)$ *of isometries of a Riemannian manifold M is a Lie transformation group with respect to the compact-open topology. For each* $x \in M$, *the isotropy subgroup* $\mathfrak{I}_x(M)$ *is compact. If M is compact,* $\mathfrak{I}(M)$ *is also compact.*

Before we begin the proof, we should perhaps point out that, *a priori*, there are two definitions of isometry for a Riemannian manifold. A diffeomorphism f of M onto itself is called an *isometry* if it preserves the metric tensor. We can also call any one-to-one mapping of M onto itself which preserves the distance function defined by the Riemannian metric an isometry of M. According to Myers and Steenrod, these two definitions are equivalent (see Kobayashi-Nomizu [1, vol. 1; p. 169] for a proof). In this book, we adopt the first definition.

Let $n = \dim M$. In the original proof of Myers and Steenrod, they took $n+1$ points $x_0, x_1, \ldots, x_n$ which are independent in a certain sense and proved that the mapping $f \in \mathfrak{I}(M) \to (f(x_0), f(x_1), \ldots, f(x_n)) \in M^{n+1} = M \times \cdots \times M$ is one-to-one and has a closed submanifold of M^{n+1} as its image. The have proved that the differentiable structure on $\mathfrak{I}(M)$ introduced by the injection $\mathfrak{I}(M) \subset M^{n+1}$ makes $\mathfrak{I}(M)$ into a Lie transformation group. Theorem 1.2 may be also derived immediately from Theorem 3.3

(Bochner-Montgomery) of Chapter I and from Theorem 1.1 (van Danzig-van der Waerden). But we prefer to derive it from Theorem 3.2 of Chapter I as follows.

Proof of Theorem 1.2. Let $L(M)$ be the bundle of linear frames over M; it is a principal bundle with group $GL(n; \mathbf{R})$, $n = \dim M$.

Lemma 1. *Let $\theta = (\theta^1, \ldots, \theta^n)$ be the canonical form on $L(M)$. For every transformation f of M, the induced automorphism $\tilde{f}$ of $L(M)$ leaves θ invariant. Conversely, every fibre-preserving transformation of $L(M)$ leaving θ invariant is induced by a transformation of M.*

Proof of Lemma 1. Let $u \in L(M)$ and $X^* \in T_u(L(M))$. We set $X = \pi(X^*) \in T_x(M)$, where $\pi\colon L(M) \to M$ is the projection and $x = \pi(u)$. Then

$$\theta(X^*) = u^{-1}(X) \quad \text{and} \quad \theta(\tilde{f} X^*) = \tilde{f}(u)^{-1}(f X),$$

where the frames u and $\tilde{f}(u)$ are considered as linear mappings of $\mathbf{R}^n$ onto $T_x(M)$ and $T_{f(x)}(M)$, respectively. It follows from the definition of $\tilde{f}$ that the following diagram is commutative:

$$\begin{array}{ccc} & \mathbf{R}^n & \\ {}^{u}\swarrow & & \searrow^{\tilde{f}(u)} \\ T_x(M) & \xrightarrow[f_*]{} & T_{f(x)}(M). \end{array}$$

Hence, $u^{-1}(X) = \tilde{f}(u)^{-1}(f X)$, thus proving that θ is invariant by $\tilde{f}$.

Conversely, if F is a fibre-preserving transformation of $L(M)$ leaving θ invariant, let f be the transformation of the base M induced by F. If we set $J = \tilde{f}^{-1} \circ F$, then J is also a fibre-preserving transformation of $L(M)$ leaving θ invariant and induces the identity transformation on the base M. Hence,

$$u^{-1}(X) = \theta(X^*) = \theta(J X^*) = J(u)^{-1}(X) \qquad \text{for } X^* \in T_u(L(M)).$$

This implies $J(u) = u$, that is, $\tilde{f}(u) = F(u)$.

Let $\omega = (\omega^i_j)_{i,j=1,\ldots,n}$ be the connection form for an affine connection of M. Then a transformation f of M is an *affine transformation* if $\tilde{f}$ preserves ω. From Lemma 1, we obtain

Lemma 2. *Let θ and ω be the canonical form and a connection form on $L(M)$ respectively. If f is an affine transformation of M, then $\tilde{f}$ preserves both θ and ω. Conversely, every fibre-preserving transformation of $L(M)$ leaving both θ and ω invariant is induced by an affine transformation of M.*

Lemma 2 implies that the group $\mathfrak{A}(M)$ of affine transformations of M is isomorphic to the group of bundle automorphisms of $L(M)$ leaving both θ and ω invariant. On the other hand, the $n + n^2$ 1-forms $\theta = (\theta^i)$ and $\omega = (\omega^i_j)$ define an absolute parallelism, i.e., a $\{1\}$-structure, on

$L(M)$. From Theorem 3.2 of Chapter I, it follows that the group $\mathfrak{A}(M)$ of affine transformations may be considered as a closed submanifold of $L(M)$. This result will be stated as Theorem 1.3 later.

Let M be a Riemannian manifold. Let $O(M)$ be the bundle of orthonormal frames over M; it is a principal bundle with group $O(n)$. We denote by θ and ω the canonical form on $O(M)$ and the connection form for the Riemannian connection, respectively. A transformation f of M is an isometry if and only if the induced bundle automorphism $\tilde{f}$ of $L(M)$ leaves $O(M)$ invariant. We denote by $\bar{f}$ the restriction of $\tilde{f}$ to $O(M)$. Since ω is the *unique* torsionfree connection in $O(M)$, every bundle automorphism of $O(M)$ leaving the canonical form θ invariant leaves the connection form ω invariant. Hence,

Lemma 3. *Every isometry f of a Riemannian manifold M induces a bundle automorphism $\bar{f}$ of $O(M)$ leaving both the canonical form θ and the connection form ω invariant. Conversely, every fibre-preserving transformation of $O(M)$ leaving both θ and ω invariant is induced by an isometry of M.*

On the other hand, $\omega=(\omega^i_j)$ is skew-symmetric, and the $\frac{1}{2}n(n+1)$ 1-forms $\theta=(\theta^i)$ and $(\omega^i_j)_{i<j}$ define an absolute parallelism on $O(M)$. By Theorem 3.2 of Chapter I, the group $\mathfrak{I}(M)$ of isometries of M may be considered as a closed submanifold of $O(M)$. An imbedding $\mathfrak{I}(M)\subset O(M)$ is defined as follows. Choose an orthonormal frame $u_0 \in O(M)$. Then an imbedding is given by

$$f\in\mathfrak{I}(M)\rightarrow\bar{f}(u_0)\in O(M).$$

Under this imbedding, the isotropy subgroup of $\mathfrak{I}(M)$ at $x_0=\pi(u_0)$ is the intersection of $\mathfrak{I}(M)$ and the fibre of $O(M)$ at x_0 in $O(M)$. Since each fibre of $O(M)$ is compact, the isotropy subgroup is also compact. If M is compact, so is the bundle space $O(M)$. Hence, its closed submanifold $\mathfrak{I}(M)$ is also compact. q.e.d.

We have proved not only Theorem 1.2 but also the following

Complement to Theorem 1.2. *The differentiable structure of $\mathfrak{I}(M)$ is given by an imbedding of $\mathfrak{I}(M)$ in the bundle $O(M)$ of orthonormal frames as a closed submanifold as follows. If u_0 is any orthonormal frame of M, then the mapping $f\in\mathfrak{I}(M)\rightarrow\bar{f}(u_0)\in O(M)$ defines an imbedding, where $\bar{f}$ is the bundle automorphism of $O(M)$ induced by f.*

In the course of the proof, we have established also the following

Theorem 1.3. *Let M be a manifold with an affine connection. Then the group $\mathfrak{A}(M)$ of affine transformations of M is a Lie transformation*

group. Its differentiable structure is given by an imbedding into the bundle $L(M)$ of linear frames as a closed submanifold as follows. If u_0 is any linear frame of M, then mapping $f \in \mathfrak{A}(M) \to \tilde{f}(u_0) \in L(M)$ defines an imbedding.

Theorem 1.3 is originally due to Nomizu [1] and Hano-Morimoto [1]. The proof given here is due to Kobayashi. Theorem 3.2 of Chapter I was proved precisely to give a unified and geometric proof for the groups of isometries, affine transformations, conformal transformations, projective transformations and, more generally, automorphisms of a Cartan connections (see Kobayashi [1]). This proof gives also an upper bound for the dimension of any of these groups. The proof of Theorem 1.2 is a special case of the proof of Theorem 5.1 of Chapter I.

2. Infinitesimal Isometries and Infinitesimal Affine Transformations

Although we are primarily interested in infinitesimal isometries here, we consider also infinitesimal affine transformations at the same time. A vector field X on a manifold with an affine connection (resp. a Riemannian manifold) is called an *infinitesimal affine transformation* (resp. *infinitesimal isometry* or *Killing vector field*) if it generates a local 1-parameter group of local affine transformations (resp. local isometries).

For any vector field X on a manifold M with an affine connection whose covariant derivation is denoted by ∇, we define a derivation A_X by

$$A_X = L_X - \nabla_X,$$

where L_X denotes the Lie derivation with respect to X. Then (cf. Kobayashi-Nomizu [1, vol. 1; p. 235])

Proposition 2.1. *For any vector fields X and Y on M, we have*

$$A_X Y = -\nabla_Y X - T(X, Y),$$

where T is the torsion tensor field of the connection ∇.

Proof. From $\nabla_X Y - \nabla_Y X - [X, Y] = T(X, Y)$ and $L_X Y = [X, Y]$, we obtain Proposition 2.1. q.e.d.

In terms of a local coordinate system $x^1, \dots, x^n$, Proposition 2.1 means that A_X is the tensor field of type (1, 1) with components

$$-\nabla_j \xi^i - \sum T^i_{kj} \xi^k,$$

where $X = \sum \xi^i \dfrac{\partial}{\partial x^i}$.

For computational purpose, the following proposition is most useful.

Proposition 2.2. (1) *A vector field X on a manifold M with an affine connection is an infinitesimal affine transformation if and only if*

$$\nabla_Y (A_X) = R(X, Y) \quad \textit{for all vector fields } Y \textit{ on } M,$$

where R is the curvature tensor. In terms of a local coordinate system,

$$\nabla_l(\nabla_j \xi^i + \sum T^i_{kj} \xi^k) + \sum R^i_{jkl} \xi^k = 0;$$

(2) *A vector field X on a Riemannian manifold M is an infinitesimal isometry if and only if A_X is skew-symmetric, i.e.,*

$$g(A_X Y, Z) + g(Y, A_X Z) = 0 \quad \text{for all vector fields } Y, Z \text{ on } M,$$

where g is the metric tensor. In terms of a local coordinate system,

$$\nabla_j \xi_i + \nabla_i \xi_j = 0.$$

Proof. (1) We prove first

Lemma 1. *A vector field X is an infinitesimal affine transformation if and only if*

$$L_X \circ \nabla_Y Z - \nabla_Y \circ L_X Z = \nabla_{[X,Y]} Z \quad \text{for all vector fields } Y, Z \text{ on } M.$$

Proof of Lemma 1. Assume that X is an infinitesimal affine transformation and let f_t be a local 1-parameter group of local transformations of M generated by X. Since f_t preserves the connection, we have

$$f_t(\nabla_Y Z) = \nabla_{f_t Y}(f_t Z) \quad \text{for all vector fields } Y, Z \text{ on } M.$$

From the definition of Lie differentiation, we obtain

$$\begin{aligned} L_X \circ \nabla_Y Z &= \lim_{t \to 0} \frac{1}{t} [\nabla_Y Z - f_t(\nabla_Y Z)] \\ &= \lim_{t \to 0} \frac{1}{t} [\nabla_Y Z - \nabla_{f_t Y} Z] + \lim_{t \to 0} \frac{1}{t} [\nabla_{f_t Y} Z - \nabla_{f_t Y}(f_t Z)] \\ &= \nabla_{L_X Y} Z + \nabla_Y \circ L_X Z = \nabla_{[X,Y]} Z + \nabla_Y \circ L_X Z. \end{aligned}$$

To prove the converse, assume the formula in Lemma. Fixing a point x of M, we set

$$V(t) = (f_t(\nabla_Y Z))_x \quad \text{and} \quad W(t) = (\nabla_{f_t Y}(f_t Z))_x.$$

For each t, both $V(t)$ and $W(t)$ are elements of $T_x(M)$. As in the proof above, we obtain

$$\frac{dV(t)}{dt} = f_t((L_X \circ \nabla_Y Z)_{f_t^{-1}(x)}),$$

$$\frac{dW(t)}{dt} = f_t(\nabla_{[X,Y]} Z + \nabla_Y \circ L_X Z)_{f_t^{-1}(x)}).$$

From our assumption, we obtain $dV(t)/dt = dW(t)/dt$. On the other hand, we have evidently $V(0) = W(0)$. Hence, $V(t) = W(t)$. This completes the proof of Lemma 1.

From Lemma 1, it follows that X is an infinitesimal affine transformations if and only if

$$L_X \circ \nabla_Y Z - \nabla_Y \circ L_X Z - (\nabla_X \circ \nabla_Y Z - \nabla_Y \circ \nabla_X Z) = \nabla_{[X,Y]} Z - [\nabla_X, \nabla_Y] Z$$

or

$$A_X \circ \nabla_Y Z - \nabla_Y \circ A_X Z = -R(X, Y) Z$$

for all vector fields Y, Z on M. But the left hand side is equal to $-(\nabla_Y(A_X)) Z$. Hence, $(\nabla_Y(A_X)) Z = R(X, Y) Z$.

(2) A vector field X is an infinitesimal isometry if and only if $L_X g = 0$.

Since g is parallel and, hence, $\nabla_X g = 0$, $L_X g = 0$ is equivalent to $A_X g = 0$. Since A_X is a derivation of the algebra of tensor fields, we have

$$A_X(g(Y, Z)) = (A_X g)(Y, Z) + g(A_X Y, Z) + g(Y, A_X Z)$$

for all vector fields Y, Z. Since A_X maps every function into zero, the left hand side vanishes. Hence, $A_X g = 0$ if and only if $g(A_X Y, Z) + g(Y, A_X Z) = 0$ for all Y, Z. q.e.d.

If X is an infinitesimal affine transformation of a Riemannian manifold M, then Proposition 2.2 implies

$$\nabla_l \nabla_j \xi^i + \sum R^i_{jkl} \xi^k = 0$$

and hence (by multiplying by g^{jl} and summing over j and l, we have)

$$\sum \nabla_j \nabla^j \xi_i + \sum R_{ij} \xi^j = 0.$$

On the other hand, applying the Laplacian Δ to the 1-form $\xi = \sum \xi_i \, dx^i$ (see the formula for Δ in Appendix 3), we obtain

$$\Delta \xi = \sum (-\nabla_j \nabla^j \xi_i + R_{ij} \xi^j) \, dx^i.$$

From these two systems of equations, we obtain

$$\Delta \xi = 2 \sum R_{ij} \xi^j \, dx^i.$$

Theorem 2.3. *Let M be a Riemannian manifold and X a vector field on M. Let ξ be the 1-form corresponding to X under the duality defined by the metric. If X is an infinitesimal isometry, it satisfies the following systems of differential equations:*

(1) $$\Delta \xi = 2 \sum R_{ij} \xi^j \, dx^i;$$

(2) $$\delta \xi = 0 \quad (\text{i.e., } \operatorname{div} X = 0).$$

Conversely, if M is compact and X satisfies (1) *and* (2), *then X is an infinitesimal isometry.*

Proof. We have shown already that if X is an infinitesimal affine transformation, it satisfies (1). If X is an infinitesimal isometry, then $\nabla_j \xi_i$ is skew symmetric in i and j (see Proposition 2.2) and its trace vanishes. This implies (2). To prove the converse, we may assume that M is orientable. (If M is not orientable, consider its orientable double covering.) We use the following integral formulas (S denoting the Ricci tensor):

$$\int_M \{S(X,X) - \mathrm{trace}(A_X \circ {}^tA_X) - \tfrac{1}{2}\mathrm{trace}((A_X + {}^tA_X)^2) - (\mathrm{div}\, X)^2\}\, dv = 0$$

and

$$\int_M \{-(\Delta X, X) + S(X,X) + \mathrm{trace}(A_X \circ {}^tA_X)\}\, dv = 0.$$

The first formula is proved in Corollary to Theorem 1 in Appendix 2. The second formula is proved in Theorem 3 of Appendix 2; we note that

$$\mathrm{trace}(A_X \circ {}^tA_X) = \sum \nabla_j \xi^i \circ \nabla^j \xi_i = (\nabla X, \nabla X).$$

By adding these two integral formulas, we obtain

$$\int_M \{-(\Delta X, X) + 2S(X,X) - \tfrac{1}{2}\mathrm{trace}((A_X + {}^tA_X)^2) - (\mathrm{div}\, X)^2\}\, dv = 0.$$

By our assumptions (1) and (2),

$$(\Delta X, X) + 2S(X,X) = 0 \quad \text{and} \quad \mathrm{div}\, X = 0.$$

(We note that ΔX is defined to be the vector field corresponding to the 1-form $\Delta \xi$; see Appendix 2.) Hence,

$$\int_M \mathrm{trace}((A_X + {}^tA_X)^2) = 0.$$

Since $A_X + {}^tA_X$ is a symmetric tensor, $\mathrm{trace}((A_X + {}^tA_X)^2)$ is the square of the length of $A_X + {}^tA_X$. It follows that $A_X + {}^tA_X = 0$. By Proposition 2.2, X is an infinitesimal isometry. q.e.d.

Theorem 2.3 and the following application is due to Yano [1].

Corollary 2.4. *Let M be a compact Riemannian manifold. Then every infinitesimal affine transformation X is an infinitesimal isometry.*

Proof. We have shown already that X satisfies (1) of Theorem 2.3. In

$$\nabla_l \nabla_j \xi^i + \sum R^i_{jkl} \xi^k = 0$$

we sum over $i = j$. Then

$$\nabla_l (\mathrm{div}\, X) = 0$$

which means that $\mathrm{div}\, X$ is a constant function on M. On the other hand,

$$\int_M (\mathrm{div}\, X)\, dv = 0.$$

Hence, $\mathrm{div}\, X = 0$, showing that (2) of Theorem 2.3 is also satisfied. Now the corollary follows from Theorem 2.3. q.e.d.

In general, an infinitesimal affine transformation or an infinitesimal isometry X generates only a local 1-parameter group of local affine transformations or local isometries. If it generates a global 1-parameter group of transformations, we call it a *complete* vector field. Thus, the group $\mathfrak{A}(M)$ of affine transformations (resp. the group $\mathfrak{I}(M)$ of isometries) of M has as its Lie algebra the set of complete infinitesimal affine transformations (resp. isometries) of M. We quote the following result (Kobayashi [4]).

Theorem 2.5. *Let M be a manifold with a complete affine connection (resp. a complete Riemannian manifold). Then every infinitesimal affine transformation (resp. isometry) is complete.*

We only sketch an outline of the proof. For detail, see Kobayashi-Nomizu [1, vol. 1; pp. 234 and 239].

Let $L(M)$ be the bundle of linear frames over M. Let $\theta=(\theta^i)$ and $\omega=(\omega^i_j)$ be the canonical form and the connection form on $L(M)$. A vector field B on $L(M)$ is called a standard horizontal vector field if $\theta^i(B)=$constant and $\omega^i_j(B)=0$. Fix a point u_0 in $L(M)$. Then for each point u of $L(M)$, there exist standard horizontal vector fields $B_1, \ldots, B_k$ and an element a of $\mathrm{GL}(n; \mathbf{R})$ such that

$$u=(b^1_{t_1} \circ b^2_{t_2} \circ \cdots \circ b^k_{t_k} u_0)\, a,$$

where each b^i_t is the 1-parameter group of transformations $\exp tB_i$ generated by B_i. Since the connection is complete, $\exp tB_i$ is defined globally. (The geodesics on M are given as the projections of the orbits of $\exp tB$ with standard horizontal B.) Let $\bar{X}$ be the infinitesimal transformation of $L(M)$ induced by X. It suffices to prove $\bar{f}_t=\exp t\bar{X}$ is defined globally since $f_t=\exp tX$ is the projection of $\bar{f}_t$. We set

$$\bar{f}_t(u)=(b^1_{t_1} \circ b^2_{t_2} \circ \cdots \circ b^k_{t_k}(\bar{f}_t(u_0)))\, a$$

for the values of t for which $\bar{f}_t(u_0)$ is defined. The fact that $\bar{f}_t(u)$ is defined independent of the choice of $B_1, \ldots, B_k$ follows from the fact that $B_1, \ldots, B_k$ are invariant by $\bar{X}$ so that $\bar{f}_t$ and $\exp tB$ commute.

On the question of extending a local (infinitesimal) isometry to a global one, see Kobayashi-Nomizu [1, vol. 1; pp. 252–256] and Nomizu [4].

3. Riemannian Manifolds with Large Group of Isometries

We first consider the following extreme case.

Theorem 3.1. *Let M be an n-dimensional Riemannian manifold. Then the group $\mathfrak{I}(M)$ of isometries is of dimension at most $\frac{1}{2}n(n+1)$. If $\dim \mathfrak{I}(M)=\frac{1}{2}n(n+1)$, then M is isometric to one of the following spaces of constant* *curvature:*

(a) *An n-dimensional Euclidean space* $\mathbf{R}^n$.

(b) *An n-dimensional sphere* S^n.

(c) *An n-dimensional projective space* $P_n(\mathbf{R})$.

(d) *An n-dimensional, simply connected hyperbolic space.*

Proof. Since $\dim O(M) = \frac{1}{2}n(n+1)$ and $\mathfrak{I}(M)$ is a closed submanifold of $O(M)$, it follows that $\dim \mathfrak{I}(M) \leqq \frac{1}{2}n(n+1)$. Suppose $\dim \mathfrak{I}(M) = \frac{1}{2}n(n+1)$. Since $\mathfrak{I}(M)$ is a closed submanifold of $O(M)$ and $\dim \mathfrak{I}(M) = \dim O(M)$, it follows that either $\mathfrak{I}(M) = O(M)$ or $\mathfrak{I}(M)$ coincides with one of the connected components of $O(M)$. (Note that $O(M)$ has one or two connected components according as M is non-orientable or orientable.) In any case, given a 2-dimensional subspace p of $T_x(M)$ and a 2-dimensional subspace p' of $T_{x'}(M)$, there is an isometry which sends p onto p'. This means that the sectional curvature determined by p coincides with the one determined by p'. This shows that M is a space of constant curvature. Since x and x' can be arbitrary points of M, we can conclude also that M is homogeneous and, hence, complete. If M is simply connected, then M must be one of (a), (b) and (d) (see, for example, Kobayashi-Nomizu [1, vol. 1; p. 265). If M is not simply connected, let $\tilde{M}$ be the universal covering manifold of M. Every infinitesimal isometry X of M induces an infinitesimal isometry $\tilde{X}$ of $\tilde{M}$ in a natural manner. Hence, $\frac{1}{2}n(n+1) = \dim \mathfrak{I}(M) \leqq \dim \mathfrak{I}(\tilde{M}) \leqq \frac{1}{2}n(n+1)$. This implies that every infinitesimal isometry $\tilde{X}$ of $\tilde{M}$ is induced by an infinitesimal isometry X of M. If we write $M = \tilde{M}/\Gamma$, where Γ is a discrete subgroup of $\mathfrak{I}(\tilde{M})$, then Γ must commute with the identity component $\mathfrak{I}^0(\tilde{M})$ of $\mathfrak{I}(\tilde{M})$. If $\tilde{M} = \mathbf{R}^n$, then $\mathfrak{I}^0(\tilde{M})$ is the group of proper motions and only the identity transformation commutes with $\mathfrak{I}^0(\tilde{M})$. If $\tilde{M} = S^n$, then $\mathfrak{I}^0(\tilde{M}) = SO(n+1)$ and only $\pm I \in O(n+1)$ commutes with $SO(n+1)$. If $\tilde{M}$ is a simply connected hyperbolic space, then $\mathfrak{I}(\tilde{M}) = O(1, n)$ (= the Lorentz group of signature $(+, -, \ldots, -)$) and $\mathfrak{I}^0(\tilde{M})$ = identity component of $O(1, n)$. In this case, the identity element is the only element of $\mathfrak{I}(\tilde{M})$ which commutes with $\mathfrak{I}^0(\tilde{M})$. Theorem 3.1 follows now immediately. q.e.d.

The fact that $\dim \mathfrak{I}(M) < \frac{1}{2}n(n+1)$ unless M has a constant curvature is classical (see, for example, Eisenhart [1]).

Theorem 3.2. *Let M be an n-dimensional Riemannian manifold with $n \neq 4$. Then the group $\mathfrak{I}(M)$ of isometries contains no closed subgroup of dimension r for $\frac{1}{2}n(n-1)+1 < r < \frac{1}{2}n(n+1)$.*

Proof. Let $\mathfrak{G}$ be a closed subgroup of dimension r of $\mathfrak{I}(M)$ and let $\mathfrak{G}_x$ denote the isotropy subgroup of $\mathfrak{G}$ at $x \in M$. Then $\mathfrak{G}_x$ is a closed subgroup of $O(n)$ by Theorem 1.1.

Lemma. *For* $n \neq 4$, $O(n)$ *contains no proper closed subgroup of dimension* $> \frac{1}{2}(n-1)(n-2)$ *other than* $SO(n)$.

Proof of Lemma. Let $\mathfrak{H}$ be a proper closed subgroup of $O(n)$. Consider $O(n)$ as a group acting on the homogeneous space $O(n)/\mathfrak{H}$. Since $O(n)$ is compact, there is an invariant Riemannian metric on $O(n)/\mathfrak{H}$. Since $O(n)$ is simple for $n \neq 4$, $O(n)$ contains no non-discrete normal subgroup and hence acts on $O(n)/\mathfrak{H}$ essentially effectively. This means that the dimension of $O(n)$ cannot exceed that of the group of isometries of $O(n)/\mathfrak{H}$. If we set $m = \dim O(n)/\mathfrak{H}$, Theorem 2.1 implies

$$\tfrac{1}{2} n(n-1) = \dim O(n) \leqq \tfrac{1}{2} m(m+1),$$

that is,

$$n \leqq m+1.$$

This implies

$$\dim \mathfrak{H} = \dim O(n) - m \leqq \tfrac{1}{2} n(n-1) - (n-1) = \tfrac{1}{2}(n-1)(n-2),$$

thus completing the proof of Lemma.

Suppose $r > \frac{1}{2} n(n-1) + 1$. Then

$$\dim \mathfrak{G}_x \geqq \dim \mathfrak{G} - \dim M > \tfrac{1}{2} n(n-1) + 1 - n = \tfrac{1}{2}(n-1)(n-2) + 1.$$

From Lemma, it follows that $\mathfrak{G}_x = O(n)$ or $\mathfrak{G}_x = SO(n)$. We shall show that $\mathfrak{G}$ is transitive on M. If x and y are two points of M which can be joined by a geodesic, let z be the midpoint of this geodesic segment and let Z be the vector tangent to the geodesic at z. Let f be a transformation belonging to $\mathfrak{G}_z$ such that $f_*(Z) = -Z$; such an isometry exists since $\mathfrak{G}_z = O(n)$ or $\mathfrak{G}_z = SO(n)$. Clearly, $f(x) = y$ and $f(y) = x$. If x and y are arbitrary points of M, we join them by a finite number of geodesic segments and apply the construction above to each segment. In this way, we see that there is an element of $\mathfrak{G}$ which sends x into y. Since $\mathfrak{G}$ is transitive on M, we have

$$r = \dim \mathfrak{G} = \dim M + \dim \mathfrak{G}_x = n + \dim O(n) = \tfrac{1}{2} n(n+1). \qquad \text{q.e.d.}$$

Theorem 3.2 is due to H.C. Wang [1]. Lemma used above is due to Montgomery and Samelson [1].

In view of Theorem 3.2, it is natural to ask which Riemannian manifolds of dimension n admits a group of isometries of dimension $\frac{1}{2} n(n-1) + 1$.

Let M be an n-dimensional Riemannian manifold with $n \neq 4$. Let $\mathfrak{G}$ be a closed subgroup of dimension $\frac{1}{2} n(n-1) + 1$ of $\mathfrak{I}(M)$. Let $\mathfrak{G}_x$ be the isotropy subgroup of $\mathfrak{G}$ at $x \in M$. We shall show that $\mathfrak{G}$ is transitive on M. Assume that it is not. Then, for every $x \in M$, the orbit of $\mathfrak{G}$ through

x is of dimension less than n. Hence,

$$\dim \mathfrak{G}_x \geqq \dim \mathfrak{G} - (n-1) = \tfrac{1}{2} n(n-1) + 1 - (n-1) = \tfrac{1}{2}(n-1)(n-2) + 1.$$

By Lemma for Theorem 2.2, either $\mathfrak{G}_x = O(n)$ or $\mathfrak{G}_x = SO(n)$. Then, as in the proof of Theorem 3.2, we see that $\mathfrak{G}$ is transitive on M. Thus, M is a homogeneous Riemannian manifold $\mathfrak{G}/\mathfrak{H}$, where $\mathfrak{H}$ is a compact group of dimension $\frac{1}{2}(n-1)(n-2)$ ($= \dim \mathfrak{G} - n$).

Lemma 1. *Let* $\mathfrak{H}$ *be a connected closed subgroup of* $SO(n)$. *If* $n \neq 4$, *then* $\mathfrak{H}$ *is isomorphic to either* $SO(n-1)$ *or the universal covering group of* $SO(n-1)$. *If* $n \neq 4, 7$, *then* $\mathfrak{H}$ *is imbedded in* $SO(n)$ *as a subgroup leaving a* 1*-dimensional subspace of* $\mathbf{R}^n$ *invariant. If* $n = 7$, *then either* $\mathfrak{H} = SO(n-1)$ *leaving a* 1*-dimensional subspace of* $\mathbf{R}^n$ *invariant or* $\mathfrak{H} = \mathrm{Spin}(7)$ *with the spin representation.*

Proof of Lemma 1. We shall prove only the first statement and indicate a proof for the remainder of Lemma 1. With respect to an invariant Riemannian metric on the homogeneous space $SO(n)/\mathfrak{H}$, the group $SO(n)$ acts as a group of isometries. Since $SO(n)$ is simple for $n \neq 4$, its action on $SO(n)/\mathfrak{H}$ is essentially effective. Since $\dim SO(n)/\mathfrak{H} = n-1$ and $\dim SO(n) = \frac{1}{2}n(n-1)$, Theorem 2.1 implies that $SO(n)$ is a maximal dimensional isometry group acting on $SO(n)/\mathfrak{H}$ and that $SO(n)/\mathfrak{H}$ is either a sphere or a real projective space. Under the linear isotropy representation, $\mathfrak{H}$ is mapped onto $SO(n-1)$. Hence, $\mathfrak{H} = SO(n-1)$ or $\mathfrak{H} = \mathrm{Spin}(n-1)$. The second and third statements tell us how $SO(n-1)$ or $\mathrm{Spin}(n-1)$ can be imbedded into $SO(n)$. The second statement is proved in Montgomery-Samelson [1] by a topological method. We indicate an algebraic proof. First, assume that the action of $\mathfrak{H}$ on $\mathbf{R}^n$ is reducible with a p-dimensional invariant subspace $\mathbf{R}^p$. Then it leaves an $(n-p)$-dimensional orthogonal complement $\mathbf{R}^{n-p}$ invariant. Hence,

$$\dim \mathfrak{H} \leqq \dim O(p) + \dim O(n-p) = \tfrac{1}{2}p(p-1) + \tfrac{1}{2}(n-p)(n-p-1).$$

This implies that $p=1$ or $p=n-1$. Next, assume that $\mathfrak{H}$ acts irreducibly on $\mathbf{R}^n$. Then $\mathfrak{H}$ is absolutely irreducible; otherwise, $\mathfrak{H}$ would be a subgroup of $U(n/2)$ of dimension $\frac{1}{4}n^2$. Now the problem is reduced to that of determining the irreducible representations of degree n of $\mathfrak{o}(n-1; \mathbf{C})$. But this can be easily accomplished by means of the theory of representations of semi-simple Lie algebras. q.e.d.

Lemma 2. *Let* $\mathfrak{G}$ *be a connected Lie group of dimension* $\frac{1}{2}n(n-1)+1$ *and* $\mathfrak{H}$ *a connected compact subgroup of* $\mathfrak{G}$ *of dimension* $\frac{1}{2}(n-1)(n-2)$ *such that its linear isotropy representation at a point of* $M = \mathfrak{G}/\mathfrak{H}$ *leaves a* 1*-dimensional subspace of the tangent space invariant. Let*

$$\mathfrak{g} = \mathfrak{h} + \mathfrak{m}' + \mathfrak{m}'' \quad \textit{(vector space direct sum)}$$

be an (ad $\mathfrak{H}$)*-invariant decomposition of the Lie algebra* $\mathfrak{g}$, *where* $\mathfrak{m}'$ *and* $\mathfrak{m}''$ *are subspaces of dimension* 1 *and* $n-1$, *respectively. Then there are the following three possibilities, provided* $n>4$:

(1) $[\mathfrak{h}, \mathfrak{m}']=0, \quad [\mathfrak{m}', \mathfrak{m}'']=0, \quad [\mathfrak{m}'', \mathfrak{m}'']=0;$

(2) $[\mathfrak{h}, \mathfrak{m}']=0, \quad [\mathfrak{m}', \mathfrak{m}'']=0, \quad [\mathfrak{m}'', \mathfrak{m}'']=\mathfrak{h};$

(3) $[\mathfrak{h}, \mathfrak{m}']=0, \quad [\mathfrak{m}', \mathfrak{m}'']=\mathfrak{m}'', \quad [\mathfrak{m}'', \mathfrak{m}'']=0$

and $[X, Y]=cY$ *for* $X\in\mathfrak{m}'$, $Y\in\mathfrak{m}''$, *where* c *is a constant which depends only on* X.

Proof of Lemma 2. Since the linear isotropy representation of $\mathfrak{H}$ is of the form

$$\begin{pmatrix} 1 & 0 \\ 0 & SO(n-1) \end{pmatrix},$$

$\mathfrak{H}$ leaves $\mathfrak{m}'$ elementwise fixed, so that $[\mathfrak{h}, \mathfrak{m}']=0$. We shall show that

$$\text{either} \quad [\mathfrak{m}', \mathfrak{m}'']=0 \quad \text{or} \quad [\mathfrak{m}', \mathfrak{m}'']=\mathfrak{m}''.$$

Fix a non-zero element X of $\mathfrak{m}'$. Since the kernel of the linear mapping $Y\in\mathfrak{m}''\to[X, Y]\in[\mathfrak{m}', \mathfrak{m}'']$ is invariant by ad $\mathfrak{H}$, it must be either 0 so that $\dim[\mathfrak{m}', \mathfrak{m}'']=\dim\mathfrak{m}''=n-1$ or the whole space $\mathfrak{m}''$ so that $[\mathfrak{m}', \mathfrak{m}'']=0$. We assume $\dim[\mathfrak{m}', \mathfrak{m}'']=n-1$. Since $\mathfrak{h}$, $\mathfrak{m}'$ and $\mathfrak{m}''$ have mutually distinct dimensions so that the irreducible representations of $\mathfrak{H}$ on $\mathfrak{h}$, $\mathfrak{m}'$ and $\mathfrak{m}''$ are mutually inequivalent, it follows that the $(n-1)$-dimensional subspace $[\mathfrak{m}', \mathfrak{m}'']$ of $\mathfrak{g}=\mathfrak{h}+\mathfrak{m}'+\mathfrak{m}''$ invariant by $\mathfrak{H}$ must coincide with $\mathfrak{m}''$. (Here, we used the assumption $n>4$.) This proves our assertion.

We shall show that if $[\mathfrak{m}', \mathfrak{m}'']=\mathfrak{m}''$, then

$$[X, Y]=cY \quad \text{for } X\in\mathfrak{m}' \text{ and } Y\in\mathfrak{m}'',$$

where c is a constant which depends only on X and not on Y. Since $X\in\mathfrak{m}'$ is invariant by $\mathfrak{H}$, the linear isomorphism $Y\in\mathfrak{m}''\to[X, Y]\in[\mathfrak{m}', \mathfrak{m}'']=\mathfrak{m}''$ commutes with the action of $\mathfrak{H}$ on $\mathfrak{m}''$. But $\mathfrak{H}$ acting on $\mathfrak{m}''$ is nothing but $SO(n-1)$. Hence, this linear isomorphism is a scalar multiple of the identity transformation.

We shall show that

$$\text{either} \quad [\mathfrak{m}'', \mathfrak{m}'']=0 \quad \text{or} \quad [\mathfrak{m}'', \mathfrak{m}'']=\mathfrak{h}.$$

Choose a unit vector $X_1\in\mathfrak{m}'$ and an orthonormal basis $X_2, \ldots, X_n$ for $\mathfrak{m}''$. Define the constants c^i_{jk} $(i, j, k=1, \ldots, n)$, by

$$[X_j, X_k]=\sum_i c^i_{jk} X_i \mod \mathfrak{h} \qquad (\text{with } c^i_{jk}=-c^i_{kj}).$$

We have to prove $c^i_{jk}=0$ for $1\leqq i\leqq n$ and $2\leqq j, k\leqq n$. Fix i, j, k. Choose an integer l, $2\leqq l\leqq n$, such that $l\neq i, j, k$. Since $n>4$, this is possible. Let A be the linear transformation of $\mathfrak{m}'+\mathfrak{m}''$ defined by

$$A(X_j)=-X_j, \quad A(X_l)=-X_l, \quad A(X_p)=X_p \quad \text{for } p\neq j, l.$$

Since A belongs to $\mathrm{SO}(n-1)$, it is induced by an element a of $\mathfrak{H}$. From

$$(\operatorname{ad} a)([X_j, X_k])=[(\operatorname{ad} a)X_j, (\operatorname{ad} a)X_k],$$

we obtain the desired relation $c^i_{jk}=0$ by comparing the coefficients of X_i on both sides. Thus, $[\mathfrak{m}'', \mathfrak{m}'']\subset\mathfrak{h}$. Since $[\mathfrak{m}'', \mathfrak{m}'']$ is an ideal of $\mathfrak{h}$ and since $\mathfrak{h}$ is simple for $n>4$, we have either $[\mathfrak{m}'', \mathfrak{m}'']=0$ or $[\mathfrak{m}'', \mathfrak{m}'']=\mathfrak{h}$.

Finally, we prove that

$$[\mathfrak{m}'', \mathfrak{m}'']=\mathfrak{h} \quad \text{implies} \quad [\mathfrak{m}', \mathfrak{m}'']=0.$$

Let $X\in\mathfrak{m}'$ and $Y, Z\in\mathfrak{m}''$ be nonzero elements such that $[Y, Z]\neq 0$. Then

$$[X,[Y,Z]]=[[X,Y],Z]+[Y,[X,Z]]=[cY,Z]+[Y,cZ]=2c[Y,Z].$$

On the other hand, from $[\mathfrak{m}', \mathfrak{h}]=0$, we obtain $[X,[Y,Z]]=0$. Hence, $c=0$. This completes the proof of Lemma 2.

We shall now consider the case where $\mathfrak{H}=\mathrm{Spin}(7)$.

Lemma 3. *Let* $\mathfrak{G}$ *be a connected Lie group of dimension* 29 $(=\frac{1}{2}8(8-1)+1)$ *and* $\mathfrak{H}=\mathrm{Spin}(7)$ *such that the linear isotropy representation at a point of* $M=\mathfrak{G}/\mathfrak{H}$ *is the spin representation. Let*

$$\mathfrak{g}=\mathfrak{h}+\mathfrak{m} \quad \textit{(vector space direct sum)}$$

be an (ad $\mathfrak{H}$)*-invariant decomposition of the Lie algebra* $\mathfrak{g}$. *Then*

$$[\mathfrak{m}, \mathfrak{m}]=0.$$

Proof of Lemma 3. Since $\dim\mathfrak{h}=21\neq 8=\dim\mathfrak{m}$, the representations of $\mathfrak{H}$ on $\mathfrak{h}$ and $\mathfrak{m}$ are mutually inequivalent. On the other hand, $[\mathfrak{m}, \mathfrak{m}]$ is an $\mathfrak{H}$-invariant subspace of $\mathfrak{g}$ and has dimension $\leqq 28$ $(=\frac{1}{2}8(8-1))$ since $\dim\mathfrak{m}=8$. Hence, we have $[\mathfrak{m}, \mathfrak{m}]=0$, $[\mathfrak{m}, \mathfrak{m}]=\mathfrak{m}$ or $[\mathfrak{m}, \mathfrak{m}]=\mathfrak{h}$.

Assume $[\mathfrak{m}, \mathfrak{m}]=\mathfrak{m}$. Let $\mathfrak{r}$ be the radical of $\mathfrak{g}$. It is invariant by $\mathfrak{H}$. On the other hand, since $\mathfrak{h}$ is simple, it follows that $\dim\mathfrak{r}\leqq\dim\mathfrak{g}-\dim\mathfrak{h}=8$. Hence, $\mathfrak{r}=\mathfrak{m}$ or $\mathfrak{r}=0$. Since $[\mathfrak{m}, \mathfrak{m}]=\mathfrak{m}$, $\mathfrak{m}$ cannot be solvable. Hence, $\mathfrak{r}=0$, i.e., $\mathfrak{g}$ is semi-simple. Since $\mathfrak{m}$ is an ideal of $\mathfrak{g}$, there is a complementary ideal $\mathfrak{h}'$. From $\dim\mathfrak{h}'=\dim\mathfrak{h}$ and from the fact that the representations of $\mathfrak{H}$ on $\mathfrak{h}$ and $\mathfrak{m}$ are inequivalent, it follows that the $\mathfrak{H}$-invariant subspace $\mathfrak{h}'$ must coincide with $\mathfrak{h}$. Hence, $[\mathfrak{h}, \mathfrak{m}]=0$. This contradicts the assumption that the linear isotropy representation of $\mathfrak{H}$ is irreducible so that $[\mathfrak{h}, \mathfrak{m}]=\mathfrak{m}$. We have thus excluded the case $[\mathfrak{m}, \mathfrak{m}]=\mathfrak{m}$.

Assume $[\mathfrak{m}, \mathfrak{m}]=\mathfrak{h}$. Let $\mathfrak{a}$ be an ideal of $\mathfrak{g}$. Since the representations of $\mathfrak{H}$ on $\mathfrak{h}$ and $\mathfrak{m}$ are inequivalent, an $\mathfrak{H}$-invariant subspace of $\mathfrak{g}$ must be $\mathfrak{g}$, $\mathfrak{h}$, $\mathfrak{m}$ or 0. Hence, $\mathfrak{a}$ must be $\mathfrak{g}$, $\mathfrak{h}$, $\mathfrak{m}$ or 0. But neither $\mathfrak{h}$ nor $\mathfrak{m}$ can be an ideal of $\mathfrak{g}$. Hence, either $\mathfrak{a}=\mathfrak{g}$ or $\mathfrak{a}=0$. This shows that $\mathfrak{g}$ is simple. But there is no simple Lie algebra of dimension 29. We have thus excluded the case $[\mathfrak{m}, \mathfrak{m}]=\mathfrak{h}$. This completes the proof of Lemma 3.

In Lemma 2 and 3, the n-dimensional Riemannian manifolds M such that $\mathfrak{I}(M)$ contains a closed subgroup of dimension $\frac{1}{2}n(n-1)+1$ have been locally determined for $n>4$. We shall now consider the global classification.

Consider first the case (1) in Lemma 2. Clearly, M is locally symmetric and flat. If M is simply connected, then $M=\mathbf{R}^n=\mathbf{R}\times\mathbf{R}^{n-1}$ and $\mathfrak{G}$ is the direct product of the group of translations on $\mathbf{R}$ and the group of proper motions of $\mathbf{R}^{n-1}$. To find a non-simply connected M, we have to look for a discrete subgroup Γ of the group of motions of $\mathbf{R}^n$ which commutes with the above $\mathfrak{G}$ elementwise. It is easy to verify that Γ must be generated by a translation of $\mathbf{R}$. In other words, if M is not simply connected, then $M=S^1\times\mathbf{R}^{n-1}$, where S^1 denotes a circle.

Consider the case (2) in Lemma 2. Clearly, M is locally symmetric and reducible. If M is simply connected, then $M=\mathbf{R}\times M''$, where M'' must be a space of constant curvature by Theorem 3.1. Since $[\mathfrak{m}'', \mathfrak{m}'']=\mathfrak{h}$, M'' is non-flat. When M is simply connected, $\mathfrak{G}$ is the direct product of the group of translations of $\mathbf{R}$ and the largest connected group of isometries of M''. This second factor is $\mathrm{SO}(n)$ or the identity component of the Lorentz group $\mathrm{O}(1, n-1)$ according as the curvature is positive or negative. It is easy to see that a discrete subgroup of $\mathfrak{I}(M)$ commuting with $\mathfrak{G}$ is generated by a translation of $\mathbf{R}$ and by $-I\in\mathrm{O}(n)$ if the curvature is positive and by a translation of $\mathbf{R}$ if the curvature is negative. In other words, if M is not simply connected, then $M=S^1\times S^{n-1}$, $M=\mathbf{R}\times P_{n-1}(\mathbf{R})$, $M=S^1\times P_{n-1}(\mathbf{R})$ or M is a product of $\mathbf{R}$ with an $(n-1)$-dimensional hyperbolic space.

We consider now the case (3) of Lemma 2. We shall show that $\mathfrak{g}$ is a subalgebra of the Lie algebra $\mathfrak{o}(1, n)$ of the Lorentz group $\mathrm{O}(1, n)$. Let

$$\mathfrak{o}(1, n)=\mathfrak{k}+\mathfrak{p}$$

be the Cartan decomposition, i.e., $\mathfrak{k}$ is the Lie algebra of a maximal compact subgroup $\mathfrak{K}$ of $\mathrm{O}(1, n)$ and $\mathfrak{p}$ is the orthogonal complement of $\mathfrak{k}$ with respect to the Killing form. The symmetric space associated with this Cartan decomposition is a hyperbolic space. Choose a $\mathfrak{K}$-invariant inner product in $\mathfrak{p}$, i.e., an invariant Riemannian metric on the associated symmetric space, in such a way that the sectional curvature is -1. Then, if X, Y and Z are three vectors in $\mathfrak{p}$, then

$$R(Y, Z)\, X=[X, [Y, Z]]=(X, Y)\, Z-(X, Z)\, Y,$$

where R is the curvature and $(\ ,\ )$ is the inner product in $\mathfrak{p}$. Choose a unit vector X in $\mathfrak{p}$ and let $\mathfrak{m}'$ be the 1-dimensional subspace of $\mathfrak{p}$ spanned by X. Let $\mathfrak{p}''$ be the orthogonal complement of $\mathfrak{m}'$ in $\mathfrak{p}$. Define a subspace $\mathfrak{m}''$ of $\mathfrak{o}(1,n)$ by

$$\mathfrak{m}'' = \{Z + [X, Z];\ Z \in \mathfrak{p}''\}.$$

Let $\mathfrak{h}$ be the subalgebra of $\mathfrak{k}$ defined by

$$\mathfrak{h} = \{W \in \mathfrak{k};\ [W, X] = 0\}.$$

It is not difficult to see that the subalgebra $\mathfrak{h} + \mathfrak{m}' + \mathfrak{m}''$ of $\mathfrak{o}(1,n)$ thus defined is isomorphic to the Lie algebra $\mathfrak{g}$ in (3) of Lemma 2. In verifying this assertion, one should choose X in Lemma 2 in such a way that the constant c is equal to 1 and also make use of the relation $[X, [Y, Z]] = (X, Y)\, Z - (X, Z)\, Y$ above.

The correspondence

$$Z \in \mathfrak{p}'' \to Z + [X, Z] \in \mathfrak{m}''$$

defines a linear isomorphism between $\mathfrak{p} = \mathfrak{m}' + \mathfrak{p}''$ and $\mathfrak{m} = \mathfrak{m}' + \mathfrak{m}''$. On the one hand, $\mathfrak{p}$ is identified with the tangent space of the symmetric space $O(1,n)/\mathfrak{K}$ at the origin and has a natural inner product $(\ ,\ )$ corresponding to the invariant Riemannian metric of curvature -1. On the other hand, $\mathfrak{m}$ can be identified with the tangent space of the homogeneous space $M = \mathfrak{G}/\mathfrak{H}$ at the origin and has an inner product $(\ ,\ )'$ which corresponds to the given Riemannian metric of M. Under the isomorphism between $\mathfrak{p}$ and $\mathfrak{m}$, these two inner products $(\ ,\ )$ and $(\ ,\ )'$ may not correspond to each other. But they are none the less closely related to each other. Since $(\ ,\ )'$ is invariant by $SO(n-1)$ (which is the subgroup of $SO(n)$ leaving the 1-dimensional subspace $\mathfrak{m}'$ of $\mathfrak{m}$ invariant), there exist positive constants a and b such that

$$(X, X)' = a(X, X) \quad \text{for } X \in \mathfrak{m}'$$

$$(Z + [X, Z], Z + [X, Z])' = b(Z, Z) \quad \text{for } Z \in \mathfrak{p}''.$$

In other words, if we define a new inner product $(\ ,\ )''$ on $\mathfrak{m}$ by

$$(X, X)'' = \frac{b}{a}(X, X)' \quad \text{for } X \in \mathfrak{m}'$$

$$(Y, Y)'' = (Y, Y)' \quad \text{for } Y \in \mathfrak{m}'',$$

then the corresponding new invariant Riemannian metric on $M = \mathfrak{G}/\mathfrak{H}$ has constant negative curvature. We claim that M is simply connected. Although this may be proved in the same way as a similar assertion in the case (2), it is an immediate consequence of the general result that a homogeneous Riemannian manifold with negative curvature must be simply connected (see Kobayashi-Nomizu [1, vol. 2; p. 105]).

Consider now the case of Lemma 3. Then M is clearly locally symmetric and flat. If M is simply connected, then $M = \mathbf{R}^8$ and $\mathfrak{G}$ is a semi-direct product of the group of translations and Spin(7) in an obvious manner. (The group of proper motions of $\mathbf{R}^8$ is a semi-direct product of the group of translations and SO(8). Since Spin(7) is a subgroup of SO(8), $\mathfrak{G}$ is a subgroup of the group of proper motions of $\mathbf{R}^8$ in a natural manner.) From the fact that Spin(7) acting on $\mathbf{R}^8$ is absolutely irreducible, it follows that the identity transformation is the only motion of $\mathbf{R}^8$ which commutes with $\mathfrak{G}$. Hence, M must be simply connected.

What we have proved may be summarized as follows.

Theorem 3.3. *Let M be an n-dimensional Riemannian manifold with $n > 4$ such that its group $\mathfrak{J}(M)$ of isometries contains a closed connected subgroup $\mathfrak{G}$ of dimension $\frac{1}{2}n(n-1)+1$. Then M must be one of the following:*

(1) *$M = \mathbf{R} \times V$, where V is a complete simply connected space of constant curvature and $\mathfrak{G} = \mathbf{R} \times \mathfrak{J}^0(V)$;*

(2) *$M = S^1 \times V$, where V is a complete simply connected space of constant curvature and $\mathfrak{G} = S^1 \times \mathfrak{J}^0(V)$;*

(3) *$M = \mathbf{R} \times P_{n-1}(\mathbf{R})$ and $\mathfrak{G} = \mathbf{R} \times \mathfrak{J}^0(P_{n-1}(\mathbf{R}))$;*

(4) *$M = S^1 \times P_{n-1}(\mathbf{R})$ and $\mathfrak{G} = S^1 \times \mathfrak{J}^0(P_{n-1}(\mathbf{R}))$;*

(5) *M is a simply connected homogeneous Riemannian manifold $\mathfrak{G}/\mathfrak{H}$ with a $\mathfrak{G}$-invariant unit vector field X and admits a $\mathfrak{G}$-invariant Riemannian metric of constant negative curvature (which agrees with the originally given metric on the tangent vectors perpendicular to X).*

If $n = 8$, then the following additional case is possible:

(6) *$M = \mathbf{R}^8$ and $\mathfrak{G} = R^8 \cdot$ Spin(7) (semi-direct product), where $\mathbf{R}^8$ denotes the translation group on $\mathbf{R}^8$ and Spin(7) is considered as a subgroup of the rotation group SO(8).*

Remark. A precise description of $\mathfrak{G}$ in (5) as a subgroup of the Lorentz group $O(1, n)$ is given in the discussion preceding the theorem. Its Lie algebra is described in (3) of Lemma 2.

A local version of Theorem 3.3 is essentially due to Yano [2] although he excluded the case $n = 8$ from consideration. The global version given here is essentially due to Kuiper [3] and Obata [1].

As we have shown above (see the paragraph preceding Lemma 1), if a four-dimensional group of isometries acts on a 3-dimensional Riemannian manifold, the action is transitive. E. Cartan ([8; pp. 293–306]) has classified all such groups together with their actions.

In the 4-dimensional case, difficulties arise from the fact that SO(4) is not simple. The 4-dimensional homogeneous Riemannian manifolds have been studied by Egorov [10] and Ishihara [1].

An extensive work on the dimension of the automorphism group of a Riemannian or affinely connected manifold has been done by Egorov [1, 15]. For a survey of the results on this subject obtained before 1956, see Yano [3]. Mann [1] has shown that if M is an n-dimensional Riemannian manifold, then $\mathfrak{I}(M)$ contains no compact subgroups of dimension r for

$$\tfrac{1}{2}(n-k)(n-k+1)+\tfrac{1}{2}k(k+1)<r<\tfrac{1}{2}(n-k+1)(n-k+2) \qquad \text{for } k=1,2,\ldots,n,$$

except for $n=4, 6, 10$; his result generalizes Theorem 3.2 when M is compact. See also Jänich [1]. For the determination of $\mathfrak{G}\subset\mathfrak{I}(M)$ of dimension $r\geqq\frac{1}{2}(n-1)(n-2)+2$ (compact or noncompact), see Kobayashi-Nagano [5] and Wakakuwa [1].

Fix a manifold M of dimension n and, for each Riemannian metric ds^2 on M, let $\mathfrak{I}(M, ds^2)$ denote the group of isometries of M with respect to ds^2. Following W.Y. Hsiang [1–3], we define the *degree of symmetry* of M to be the maximum of dim $\mathfrak{I}(M, ds^2)$ for all possible Riemannian metrics ds^2 of M. It is a non-negative integer not exceeding $\frac{1}{2}n(n+1)$. Define also the *degree of compact symmetry* of M to be the maximum dimension of all possible compact Lie groups acting on M. If $\mathfrak{G}$ is a compact Lie group acting on M, there is a $\mathfrak{G}$-invariant Riemannian metric on M. Hence, the degree of compact symmetry of M does not exceed the degree of symmetry of M. If M is compact, $\mathfrak{I}(M, ds^2)$ is compact and hence the two degrees coincide. The results in this section may be interpreted in terms of these degrees. See also Ku-Mann-Sicks [1] on the degree of symmetry.

4. Riemannian Manifolds with Little Isometries

In the preceding section, we considered Riemannian manifolds which admit many isometries. We consider here those which admit hardly any isometries. We begin with the following theorem of Bochner [1].

Theorem 4.1. *Let M be a compact Riemannian manifold with negative Ricci tensor. Then the group $\mathfrak{I}(M)$ of isometries is finite.*

Proof. Since $\mathfrak{I}(M)$ is compact by Theorem 1.2, it suffices to show that dim $\mathfrak{I}(M)=0$, i.e., M admits no infinitesimal isometries. Let X be a vector field on M. We make use of the tensor field A_X of type (1, 1) defined by $A_X=L_X-\nabla_X$ as a derivation in § 2. Since the Riemannian connection is torsionfree, we have $A_X Y = -\nabla_Y X$ for all vector fields Y, i.e., $A_X = -\nabla X$ (see Proposition 2.1). We proved (in Proposition 2.2) that X is an infinitesimal affine transformation if and only if

$$\nabla_Y(A_X) = R(X, Y) \qquad \text{for all vector fields } Y.$$

Lemma 1. *Let x be a point of M and $V_1, \ldots, V_n$ an orthonormal basis for the tangent space $T_x(M)$. If X is an infinitesimal affine transformation of M, then*

$$g\left(\sum_{i=1}^{n} (\nabla_{V_i} A_X) V_i, X\right) = S(X, X),$$

where S denotes the Ricci tensor and g the metric tensor.

Proof of Lemma 1. Since $\nabla_{V_i} A_X = R(X, V_i)$, we have

$$g\left(\sum_{i=1}^{n} (\nabla_{V_i} A_X) V_i, X\right) = \sum_{i=1}^{n} g\big(R(X, V_i) V_i, X\big) = S(X, X),$$

thus proving Lemma 1.

If f is a function on M, its second covariant derivative $\nabla^2 f$ may be considered as a covariant symmetric tensor field of degree 2 and defines a symmetric bilinear forms on each tangent space. If $V_1, \ldots, V_n$ is an orthonormal basis for $T_x(M)$ as above, then the Laplacian $(\Delta f)_x$ of f at x is given by

$$(\Delta f)_x = \sum_{i=1}^{n} \nabla^2 f(V_i, V_i).$$

Lemma 2. *Let X be an infinitesimal affine transformation of M and f the function on M defined by $f = \frac{1}{2} g(X, X)$. Then*

(1) $\quad \nabla^2 f(V, V) = g(\nabla_V X, \nabla_V X) - g\big(R(X, V) V, X\big) \quad$ *for* $V \in T_x(M)$;

(2) $\quad (\Delta f)_x = \sum_{i=1}^{n} g(\nabla_{V_i} X, \nabla_{V_i} X) - S(X, X),$

where $V_1, \ldots, V_n$ is an orthonormal basis for $T_x(M)$.

Proof of Lemma 2. We extend each V to a vector field in a neighborhood of x in such a way that $\nabla_V V = 0$ at x. This can be accomplished by parallel displacing V along the geodesic $\exp(tV)$ for small values t and then extending it around the geodesic segment. Then

$$\begin{aligned}
\nabla^2 f(V, V)_x = \big(V(Vf)\big)_x &= g(\nabla_V X, \nabla_V X)_x + g(\nabla_V \nabla_V X, X)_x \\
&= g(\nabla_V X, \nabla_V X)_x - g\big(\nabla_V(A_X V), X\big)_x \\
&= g(\nabla_V X, \nabla_V X)_x - g\big((\nabla_V A_X) V, X\big)_x - g\big(A_X(\nabla_V V), X\big)_x \\
&= g(\nabla_V X, \nabla_V X)_x - g\big(R(X, V) V, X\big)_x,
\end{aligned}$$

thus proving (1). Applying (1) to $V=V_i$ and summing the resulting equalities with respect to i, we obtain (2).

To complete the proof of Theorem 4.1, we apply Green's theorem

$$\int_M \Delta f \cdot dv = 0$$

to the function $f=\frac{1}{2}g(X, X)$. On the other hand, if X is an infinitesimal affine transformation, then $\Delta f > 0$ by (2) of Lemma 2 unless X is identically zero. In applying Green's theorem, we have to assume that M is orientable. If M is not orientable, we have only to consider an orientable double covering of M. q.e.d.

From the proof above we obtain also the following result.

Corollary 4.2. *If M is a compact Riemannian manifold with negative semi-definite Ricci tensor, then every infinitesimal isometry of M is a parallel vector field.*

Theorem 4.1 may be generalized to a non-compact manifold to some extent.

Theorem 4.3. *Let M be a Riemannian manifold with negative definite Ricci tensor. If the length of an infinitesimal isometry X attains a local maximum at some point of M, then X vanishes identically on M.*

Proof. Suppose $f=\frac{1}{2}g(X, X)$ attains a local maximum at x. Then $(\Delta f)_x \leqq 0$. On the other hand, (2) of Lemma 2 above implies that $\Delta f > 0$ wherever $X \neq 0$. Hence, X must vanish at x. Since f attains a local maximum at x, this means that X vanishes in a neighborhood of x. By applying "Complement to Theorem 1.2" to the local 1-parameter group generated by X, we see that X vanishes everywhere on M. q.e.d.

The following result is due to Frankel [3]. With a stronger assumption than in Theorem 4.1, we can prove a little more.

Theorem 4.4. *Let M be a compact Riemannian manifold with non-positive sectional curvature and with negative definite Ricci tensor. An isometry f of M which is homotopic to the identity transformation must be the identity transformation.*

Proof. Let h_t, $0 \leqq t \leqq 1$, be a homotopy such that h_0 is the identity transformation and $h_1 = f$. Let $\tilde{M}$ be the universal covering space of M with covering projection $\pi: \tilde{M} \to M$. Let $\tilde{h}_t$ be the unique lift of h_t such that $\tilde{h}_0$ is the identity transformation of $\tilde{M}$. (By a lift of h_t, we mean $\tilde{h}_t: \tilde{M} \to \tilde{M}$ such that $\pi \circ \tilde{h}_t = h_t \circ \pi$.) We set $\tilde{f} = \tilde{h}_1$. Then $\tilde{f}$ is a lift of f and, hence, is an isometry of $\tilde{M}$. Since each transformation $\tilde{h}_t$ normalizes the group of deck-transformations which is a discrete group, the 1-parameter family $\tilde{h}_t$ of transformations must commute with the deck-transformations elementwise. Since the deck-transformations are all isometries of $\tilde{M}$, it follows that, for every $\tilde{p} \in \tilde{M}$, the distance $d(\tilde{p}, \tilde{f}(\tilde{p}))$ between $\tilde{p}$ and $\tilde{f}(\tilde{p})$ depends only on $\pi(\tilde{p})$. Hence, since M is compact, $d(\tilde{p}, \tilde{f}(\tilde{p}))$ attains an absolute, hence relative maximum at some point $\tilde{p}_0$ of $\tilde{M}$. We wish to calculate the Hessian and then the Laplacian of this function $d(\tilde{p}, \tilde{f}(\tilde{p}))$ at $\tilde{p}_0$.

Let c be the geodesic from $\tilde{p}_0$ to $\tilde{f}(\tilde{p}_0)$; since $\tilde{M}$ is simply connected and complete with non-positive sectional curvature, this geodesic exists and is unique (see, for example, Kobayashi-Nomizu [1; vol. 2; p. 102]). For each unit tangent vector X_0 at $\tilde{p}_0$ perpendicular to the geodesic c, we define a Jacobi field X along c extending X_0 as follows. Consider the geodesic $\exp(sX_0)$, $|s|<\varepsilon$, through $\tilde{p}_0$. For each fixed s, let c_s be the geodesic joining the point $\exp(sX_0)$ to the point $f(\exp(sX_0))$. Then c_s, $|s|<\varepsilon$, is a 1-parameter family variation of the geodesic segment c. Let X be the infinitesimal variation, i.e., the Jacobi field, defined by the variation c_s (cf. Kobayashi-Nomizu [1; vol. 2; p. 63]). Denote by T the unit vector field tangent to the geodesic c, i.e., the velocity vector field of c. The formula for the first variation of arc-length is given by (see, for example, Kobayashi-Nomizu [1, vol. 2; p. 80])

$$dL(X) = g(X, T)_{\tilde{f}(\tilde{p}_0)} - g(X, T)_{\tilde{p}_0}.$$

Since the length $L(s)$ of c_s is the distance between $\exp(sX_0)$ and $f(\exp(sX_0))$ and c is the longest curve in the family c_s by hypothesis, we have $dL(X)=0$. Since $g(X, T)_{\tilde{p}_0}=0$, it follows that $g(X, T)_{\tilde{f}(\tilde{p}_0)}=0$. Since the Jacobi field X is perpendicular to T at two points, it is perpendicular to T everywhere along c. The second variation $I(X, X) = (d^2L/ds^2)_{s=0}$ is given by (cf. Kobayashi-Nomizu [1, vol. 2; p. 81])

$$I(X, X) = \int_0^a \big(g(X', X') - g(R(X, T)\,T, X)\big)\,dt \qquad (\text{where } X' = \nabla_T X),$$

where t denotes the arc-length parameter for c which is parametrized by $0 \leqq t \leqq a$.

Let $V_1, \ldots, V_{n-1}$ be tangent vectors at $\tilde{p}_0$ such that $T, V_1, \ldots, V_{n-1}$ form an orthonormal basis of $T_{\tilde{p}_0}(\tilde{M})$. To each V_i, we apply the construction above to obtain a Jacobi field, denoted also by V_i, along c. Then

$$\sum_{i=1}^{n-1} I(V_i, V_i) \geqq -\int_0^a \sum_{i=1}^{n-1} g(R(V_i, T)\,T, V_i)\,dt.$$

By our assumption on the sectional curvature the integrand is non-negative along c. If we denote by S the Ricci tensor as before, then the integrand coincides with $S(T, T)_{\tilde{p}_0}$ at $\tilde{p}_0$ since $T, V_1, \ldots, V_{n-1}$ is an orthonormal basis for $T_{\tilde{p}_0}(\tilde{M})$. Hence, $\sum_{i=1}^{n-1} I(V_i, V_i) > 0$ if c has positive length.

Thus, if c has positive length, then $I(V_i, V_i) > 0$ for some V_i, which contradicts the fact that c is the longest of the curves. Hence, c must have zero length, i.e., $\tilde{f}(\tilde{p}_0) = \tilde{p}_0$. Since the distance $(\tilde{p}, \tilde{f}(\tilde{p}))$ attains a relative maximum (actually an absolute maximum) at $\tilde{p}_0$, it follows that $\tilde{f}(\tilde{p}) = \tilde{p}$ in a neighborhood of $\tilde{p}_0$ and hence everywhere on M. q.e.d.

We have assumed that in the homotopy h_t each h_t is a diffeomorphism of M onto itself. The full strength of compactness of M was not used in the proof; the theorem still holds if the function $F(p) = \text{distance}$ $(\tilde{p}, \tilde{f}(\tilde{p}))$, where $p = \pi(\tilde{p})$, attains a relative maximum at some point of M. The function F on M plays the same role as the length of an infinitesimal isometry X in the proof of Theorem 4.1 or 4.3. This function has been systematically exploited by Ozols [1].

For non-differentiable versions of some of the results in this section, see Busemann [2].

The following result of Atiyah-Hirzebruch [1] implies that a compact manifold M with nonzero $\hat{A}$-genus cannot admit a Riemannian metric for which $\dim \mathfrak{J}(M) > 0$, i.e., its degree of symmetry is zero.

Theorem 4.5. *If a circle group acts differentiably on a compact manifold M, then the $\hat{A}$-genus of M vanishes.*

5. Fixed Points of Isometries

The following elementary result shows that the fixed point set of a family of isometries is a nice differential geometric object.

Theorem 5.1. *Let M be a Riemannian manifold and $\mathfrak{G}$ any set of isometries of M. Let F be the set of points of M which are left fixed by all elements of $\mathfrak{G}$. Then each connected component of F is a closed totally geodesic submanifold of M.*

Proof. Assuming that F is non-empty, let x be a point of F. Let V be the subspace of $T_x(M)$ consisting of vectors which are left fixed by all elements of $\mathfrak{G}$. Let U^* be a neighborhood of the origin in $T_x(M)$ such that the exponential mapping $\exp_x\colon U^* \to M$ is an injective diffeomorphism. Let $U = \exp_x(U^*)$. We may further assume that U is a convex neighborhood. Then it is easy to see that $U \cap F = \exp_x(U^* \cap V)$. This shows that a neighborhood $U \cap F$ of x in F is a submanifold $\exp_x(U^* \cap V)$. Hence F consists of submanifolds of M. It is clear that F is closed. If two points of F are sufficiently close so that they can be joined by a unique minimizing geodesic, then every point of this geodesic must be fixed by $\mathfrak{G}$. Hence, each component of F is totally geodesic. q.e.d.

Remark. More generally, if M is a manifold with an affine connection and $\mathfrak{G}$ is a set of affine transformations of M, then the set F of points left fixed by $\mathfrak{G}$ is a disjoint union of closed, auto-parallel submanifolds (see Kobayashi-Nomizu [1, vol. 2; p. 61]).

The following result shows that the number of connected components in F is limited.

Corollary 5.2. *In Theorem* 5.1, *assume that M is complete. Let p and q be points belonging to different components of F. Then q is a cut point of p. If $\mathfrak{G}$ is a connected Lie group of isometries, then q is a conjugate point of p.*

Proof. If q is not a cut point, then by definition there is a unique minimizing geodesic, say c, from p to q. If f is any element of $\mathfrak{G}$, then $f(c)$ is also geodesic from p to q with the same arc-length as c. Hence $c=f(c)$ and every point of c is left fixed by f. Since f is an arbitrary point of $\mathfrak{G}$, this shows that c is contained in F and, hence, p and q are in the same connected component of F.

Let $\mathfrak{G}$ be a connected Lie group with positive dimension. Assume that q is not conjugate to p. Let c be a geodesic from p to q. Every infinitesimal isometry X defines a Jacobi field along c; the one-parameter group generated by X defines a variation of c in a natural manner. If X belongs to the Lie algebra of $\mathfrak{G}$, then X vanishes at p and q. Since q is not conjugate to p, X must vanish at every point of c. Thus, the 1-parameter subgroup of $\mathfrak{G}$ generated by X leaves c fixed pointwise. Since $\mathfrak{G}$ is connected and is generated by these 1-parameter subgroups, $\mathfrak{G}$ leaves c fixed. This shows that p and q can be joined by a curve c contained in F. q.e.d.

Theorem 5.1 can be strengthened if $\mathfrak{G}$ is a 1-parameter group. Indeed, we have the following result (Kobayashi [5]).

Theorem 5.3. *Let M be a Riemannian manifold and X an infinitesimal isometry of M. Denote by* $\mathrm{Zero}(X)$ *the set of points of M where X vanishes. Let* $\mathrm{Zero}(X)=\bigcup_i N_i$ *be the decomposition of* $\mathrm{Zero}(X)$ *into its connected components. Then:*

(1) *Each N_i is a closed totally geodesic submanifold of even codimension.*

(2) *Considered as a field of linear endomorphisms of $T(M)$, the covariant derivative $\nabla X(=-A_X)$ annihilates the tangent bundle $T(N_i)$ of each N_i and induces a (skew-symmetric) automorphism of the normal bundle $T^{\perp}(N_i)$ of each N_i. Restricted to each N_i, ∇X is parallel, $\nabla_V(\nabla X)=0$ for every $V\in T(N_i)$.*

(3) *The normal bundle $T^{\perp}(N_i)$ of N_i can be made into a complex vector bundle.*

(4) *If M is orientable, then each N_i is orientable.*

Proof. Applying the proof of Theorem 5.1 to the local 1-parameter group of local isometries, we see that each N_i is a closed totally geodesic submanifold. Let $x \in N_i$. If we choose a suitable orthonormal basis for $T_x(M)$, then the linear endomorphism $(\nabla X)_x$ of $T_x(M)$ is given by a matrix of the form

$$A = \begin{pmatrix} 0 & & & & & & & \\ & \ddots & & & & & & \\ & & 0 & & & & & \\ & & & 0 & a_1 & & & \\ & & & -a_1 & 0 & & & \\ & & & & & \ddots & & \\ & & & & & & 0 & a_k \\ & & & & & & -a_k & 0 \end{pmatrix}, \qquad a_i \neq 0.$$

Write $T_x(M) = T'_x + T''_x$, where T'_x (resp. T''_x) is the subspace spanned by the first $n-2k$ elements (resp. the last $2k$ elements) of the basis of $T_x(M)$. Then T'_x is left fixed pointwise by $\exp(tA)$ and hence by $\exp(tX)$. It is clear that T'_x is the tangent space $T_x(N_i)$ and T''_x is the normal space $T_x^\perp(N_i)$. This proves (1) and the first statement of (2). According to Proposition 2.2, we have $\nabla_Y(A_X) = R(X, Y)$ for every vector Y on M. This may be rewritten as

$$\nabla_Y(\nabla X) = -R(X, Y) \quad \text{for } Y \in T(M).$$

At every point of Zero(X), the right hand side $R(X, Y)$ vanishes and hence $\nabla_Y(\nabla X)$ for every vector tangent to M at a point of Zero(X). This proves a little more than what is claimed in (2).

Since the eigen-values $\pm i a_1, \ldots, \pm a_k$ of $(\nabla X)_x$ defined above remain constant on each N_i because ∇X is parallel on N_i, we can decompose the normal bundle $T^\perp(N_i)$ into subbundles $E_1, \ldots, E_r$ as follows:

$$T^\perp(N_i) = E_1 + \cdots + E_r \qquad \text{(orthogonal decomposition)},$$

where $E_1, \ldots, E_r$ correspond to the eigen-values $-b_1^2, \ldots, -b_r^2$ of $(\nabla X)_x$ restricted to $T^\perp(N_i)$. Let J be the endomorphism of $T^\perp(N_i)$ defined by $J|E_j = \frac{1}{b_j}(\nabla X)$. Then $J^2 = -I$, and J defines a complex vector bundle structure in $T^\perp(N_i)$. Since $T(M)|N_i = T(N_i) + T^\perp(N_i)$, if $T(M)$ is orientable, $T(N_i)$ is also orientable. q.e.d.

Since the fixed point set F of isometries or the zero set of infinitesimal isometries consists of totally geodesic submanifolds, it is perhaps appropriate to mention the following result on totally geodesic submanifolds at this point.

Theorem 5.4. *Let N be a totally geodesic submanifold of a Riemannian manifold M. If X is an infinitesimal isometry of M, its restriction to N projected upon N defines an infinitesimal isometry of N. In particular, if M is homogeneous and N is a closed totally geodesic submanifold, then N is also homogeneous.*

For the proof, see Kobayashi-Nomizu [1, vol. 2; p. 59], where other properties of totally geodesic submanifolds are also given.

This is probably an appropriate place to mention the following topological result.

Theorem 5.5. *Let M be a compact Riemannian manifold and X be an infinitesimal isometry of M. Let* $\mathrm{Zero}(X)=\bigcup N_i$ *be the decomposition of the zero set of X into its connected components. Then*

(1) $\sum_k (-1)^k \dim H_k(M; K) = \sum_i (\sum_k (-1)^k \dim H_k(N_i; K))$,

(2) $\sum_k \dim H_k(M; K) \geqq \sum_i (\sum_k \dim H_k(N_i; K))$ *for any coefficient field K.*

Proof. We shall prove (1) following Kobayashi [5]. The proof of (2) is harder and is omitted (see Floyd [1] and Conner [1]).

Let A_i be the closure of an ε-neighborhood of N_i. We take ε so small that every point of A_i can be joined to the nearest point of N_i by a unique geodesic of length $\leqq \varepsilon$ and that $A_i \cap A_j$ be empty for $i \neq j$. Then A_i is a fibre bundle over N_i whose fibres are closed solid balls of radius ε. Set $A=\bigcup A_i$. Let B be the closure of the open set $M-A$. Then $A\cap B$ is the boundary of A.

We remark that if

$$\cdots \to U_k \to V_k \to W_k \to U_{k-1} \to V_{k-1} \to \cdots$$

is an exact sequence of vector spaces, then

$$\sum (-1)^k \dim U_k - \sum (-1)^k \dim V_k + \sum (-1)^k \dim W_k = 0.$$

We apply this formula to the exact sequences of homology groups (with coefficient field K) induced by

$$B \to M \to (M, B) \quad \text{and} \quad A\cap B \to A \to (A, A\cap B)$$

and obtain

$$\chi(B) - \chi(M) + \chi(M, B) = 0 \quad \text{and} \quad \chi(A\cap B) - \chi(A) + \chi(A, A\cap B) = 0,$$

where χ denotes the Euler number. By Excision Axiom, (M, B) and $(A, A\cap B)$ have the same relative homology. Hence, $\chi(M, B) = \chi(A, A\cap B)$. It follows that

$$\chi(M) = \chi(A) + \chi(B) - \chi(A\cap B).$$

Since the 1-parameter group generated by X has no fixed points in B nor in $A \cap B$, Lefschetz Theorem implies $\chi(B) = \chi(A \cap B) = 0$. Hence, $\chi(M) = \chi(A)$. Since A_i is a fibre bundle over N_i with solid ball as fibre, we have

$$\chi(A_i) = \chi(N_i).$$

Finally, we obtain

$$\chi(M) = \sum \chi(A_i) = \sum \chi(N_i).$$

We should remark that in Floyd [1] and Conner [1] (2) is stated as follows. If T is a total group acting on a manifold M with fixed point set F, then for any coefficient field K we have

$$\sum \dim H_k(M; K) \geqq \sum \dim H_k(F; K).$$

To see that their statement means (2), let T be the closure of the 1-parameter subgroup of $\mathfrak{I}(M)$ generated by X. Then T is a connected compact abelian group and hence a toral group. Clearly, $F = \mathrm{Zero}(X)$. q.e.d.

As a generalization of (1) of Theorem 5.5, we mention the following result. *Let M be a compact Riemannian manifold and f be an isometry of M. Let F be the fixed point set of f. If we denote the Lefschetz number of f by $L(f)$ and the Euler number of F by $\chi(F)$, then*

$$L(f) = \chi(F).$$

To prove this statement, we have only to replace $\mathrm{Zero}(X)$ by F and the Euler numbers $\chi(M)$, $\chi(A)$, ... by the Lefschetz numbers $L(f)$, $L(f|A)$, ... This result has been proved by Huang [1] when f is periodic.

We shall see that (1) of Theorem 5.5 is a very special case of Theorem 6.1 in the next section. It becomes also a special case of the classical theorem of Hopf when the zeros of X are isolated since an infinitesimal isometry has index 1 at each of its isolated zeros.

For various homological results on periodic transformations and toral group actions, see Borel [3] and the references therein.

An infinitesimal version of the following theorem is due to Berger [1]. The generalization given here is due to Weinstein [1]; he proved the result for a conformal transformation. The idea of the proof is similar to that of Frankel [2].

Theorem 5.6. *Let M be a compact, orientable Riemannian manifold with positive sectional curvature. Let f be an isometry of M.*

(1) *If $n = \dim M$ is even and f is orientation-preserving, then f has a fixed point.*

(2) *If $n = \dim M$ is odd and f is orientation-reversing, then f has a fixed point.*

Proof. Let $d(p, f(p))$ be the distance between $p \in M$ and $f(p) \in M$. It is a non-negative function on M. Let p_0 be a point of M where this function $d(p, f(p))$ achieves a minimum. We must show that the minimum is zero. Assume that $d(p_0, f(p_0))$ is positive. Let c be a minimizing geodesic from p_0 to $f(p_0)$.

Let N and N' be the normal space to c at p_0 and $f(p_0)$, respectively. We shall show that f maps N onto N'. We consider a 1-parameter family of curves c_s, $-\varepsilon < s < \varepsilon$, such that $c = c_0$ and, for each fixed s, the starting point of c_s is mapped into the end point of c_s by f. If we denote by $L(s)$ the arc-length of c_s, then $L'(0) = 0$. On the other hand, (see for instance Kobayashi-Nomizu [1, vol. 2; p. 80])

$$L'(0) = g(X, T)_{f(p_0)} - g(X, T)_{p_0} - \int_0^a g(X, \nabla_T T)\, dt,$$

where T is the vector field tangent to c, X is the variation vector field defined by the family c_s and a is the arc-length of c. Since c is a geodesic, the integrand on the right hand side vanishes. The vector field X can be prescribed at p_0. In particular, let X be perpendicular to T at p_0. Then $L'(0) = 0$ implies that X is perpendicular to T at $f(p_0)$. Since $X_{f(p_0)} = f(X_{p_0})$, this proves that f maps N onto N'.

We claim that f maps the initial velocity vector T_{p_0} of c to the velocity vector $T_{f(p_0)}$ of c at $f(p_0)$. Since f maps N onto N', it is clear that $f(T_{p_0})$ is either $T_{f(p_0)}$ or $-T_{f(p_0)}$. If we choose X in the formula above for $L'(0)$ in such a way that $X_{p_0} = T_{p_0}$, then $g(X, T)_{f(p_0)} = g(X, T)_{p_0} = 1$, which implies our assertion.

If we denote by A the composition of f: $T_{p_0}(M) \to T_{f(p_0)}(M)$ and the parallel displacement from $f(p_0)$ to p_0 along c (in the reversed direction), then we have a linear automorphism of the tangent space $T_{p_0}(M)$ leaving N invariant. According as f is orientation-preserving or orientation-reversing, the orthogonal transformation A (and also its restriction to N) has determinant 1 or -1. If $n = \dim M$ is even (resp. odd) and f is orientation-preserving (resp. orientation-reversing), then A leaves a unit vector, say X, of N fixed. We extend this vector to a parallel vector field X along c by parallel displacement. By construction, $f(X_{p_0}) = X_{f(p_0)}$. We define a 1-parameter family of curves c_s, $-\varepsilon < s < \varepsilon$, as follows. For each fixed t, we set $c_s(t) = \exp(s X_{c(t)})$. Then for each fixed t, $c_s(t)$ describes a geodesic as s varies and has $X_{c(t)}$ as the tangent vector at $s = 0$. Let a be the arc-length of c so that $f(p_0) = c(a)$. Since $f(X_{c(0)}) = X_{c(a)}$, it follows that $f(c_s(0)) = c_s(a)$. Let $L(s)$ denote the arc-length of c_s. From the way p_0 was chosen, it is clear that $L(s)$ achieves a minimum at $s = 0$. Hence, $L''(0) \geqq 0$. On the other hand, the second variation $L''(0)$ of the arc-length is given by

$$L''(0) = \int_0^a [g(\nabla_T X, \nabla_T X) - g(R(X, T)\, T, X)]\, dt.$$

(In general, terms involving the second fundamental forms of the 1-dimensional submanifolds $c_s(0)$ and $c_s(a)$ must be also taken into account. But they do not appear here since both $c_s(0)$ and $c_s(a)$ are geodesics. See, for instance, Bishop-Crittenden [1; p. 219].)

Since X is parallel, we have $\nabla_T X=0$. By the hypothesis on the curvature, we have $g(R(X, T)\, T, X)>0$. Hence, the formula above implies $L''(0)<0$. This is a contradiction. q.e.d.

From Theorem 5.6 we may derive the following result of Berger [1].

Corollary 5.7. *If M is an even dimensional, compact Riemannian manifold with positive sectional curvature, then every infinitesimal isometry X of M has a zero.*

Proof. If X has no zeros, then $\exp(tX)$ has no fixed points for small t. If M is orientable, this is a contradiction by Theorem 5.6. If M is not orientable, let $\tilde{M}$ be the orientable double covering space of M and let $\tilde{X}$ be the infinitesimal isometry of $\tilde{M}$ induced by X. Then apply the same reasoning to $\tilde{M}$ and $\tilde{X}$. q.e.d.

Remark. The original proof of Berger goes as follows. Set $f=\frac{1}{2}g(X, X)$ and let p_0 be a point where f achieves a minimum. If V is a non-zero vector at p_0, then $\nabla^2 f(V, V)\geqq 0$; for the second covariant derivative $\nabla^2 f$ of at p_0 is the Hessian of f at p_0. On the other hand, $\nabla^2 f(V, V)$ is given by (Lemma 2 for Theorem 4.1)

$$\nabla^2 f(V, V)=g(\nabla_V X, \nabla_V X)-g(R(X, V)\, V, X).$$

Assume that X has no zeros, i.e., f is positive everywhere. We shall find a vector V at p_0 such that $\nabla^2 f(V, V)$ is negative. Consider the linear endomorphism of the tangent space $T_{p_0}(M)$ given by ∇X $(=-A_X)$. We claim that ∇X annihilates X at p_0, i.e., $(\nabla_X X)_{p_0}=0$. In fact, for every vector U at p_0, we have $0=Uf=g(\nabla_U X, X)=-g(\nabla_X X, U)$, the last equality being the consequence of the skew-symmetricity of ∇X. Hence, $(\nabla_X X)_{p_0}=0$. Being a skew-symmetric linear endomorphism of $T_{p_0}(M)$, ∇X is of even rank. Since it annihilates X_{p_0}, it has to annihilate another nonzero vector, say V, at p_0 perpendicular to X. Then $\nabla_V X=0$. Since the sectional curvature $g(R(X, V)\, V, X)$ is positive, it follows that $\nabla^2 f(V, V)$ is negative. This is a contradiction.

Corollary 5.8. *Let M be a compact Riemannian manifold with positive sectional curvature.*

(1) *If* $\dim M$ *is even and M is orientable, then M is simply connected.*

(2) *If* $\dim M$ *is odd, then M is orientable.*

Proof. (1) Let $\tilde{M}$ be the universal covering space of M. Every deck-transformation of $\tilde{M}$ is orientation-preserving and hence must have a fixed point by Theorem 5.6. This is a contradiction unless M itself is simply connected.

(2) If M is not orientable, let $\tilde{M}$ be the orientable double covering space of M. Then the non-trivial deck-transformation of $\tilde{M}$ is orientation-reversing and must have a fixed point by Theorem 5.6. This is a contradiction. q.e.d.

The corollary above is originally due to Synge [1]. From this corollary and Theorem 5.6, we obtain the following result.

Corollary 5.9. *Let M be a compact Riemannian manifold with positive sectional curvature. If M is not orientable, then every isometry of M has a fixed point.*

Proof. By Corollary 5.8, $\dim M$ is even, and the orientable double covering space $\tilde{M}$ of M is simply connected. We lift an isometry f of M to an isometry $\tilde{f}$ of $\tilde{M}$. By composing it with the non-trivial deck-transformation if necessary, we may assume that $\tilde{f}$ is orientation-preserving. By Theorem 5.6, $\tilde{f}$ has a fixed point. Hence, f has a fixed point. q.e.d.

For other applications of Theorem 5.6, see Weinstein [1].

In the case of non-positive curvature, the following theorem of E. Cartan is basic (see Kobayashi-Nomizu [1, vol. 2; p. 111] for a proof):

Theorem 5.10. *Every compact group $\mathfrak{G}$ of isometries of a complete, simply connected Riemannian manifold M with non-positive sectional curvature has a fixed point.*

We should point out that the fixed point set F in Theorem 5.10 is connected. Suppose p and q are two points of F and consider the (unique) geodesic from p to q. Every element of $\mathfrak{G}$ leaves this geodesic pointwise fixed since it leaves p and q fixed. This shows that p and q can be joined by a geodesic which lies in F.

In connection with this, we note that if $\mathfrak{G}$ is a *connected* Lie group of isometries acting on a complete Riemannian manifold with non-positive positive sectional curvature (not necessarily simply connected), then its fixed point set F is connected (possibly empty, of course). This follows from Corollary 5.2 and from the fact that M is free of conjugate points.

Let M be a complete Riemannian manifold with non-positive sectional curvature and $\pi_1(M)$ be its fundamental group. Then each element of $\pi_1(M)$ acts on the universal covering manifold $\tilde{M}$ without fixed point. Hence, the study of the fixed points of an isometry of $\tilde{M}$ has a bearing on the study of $\pi_1(M)$. In this connection, see Preismann [1], Bishop-O'Neill [1], Wolf [4], Yau [1], Gromoll-Wolf [1].

6. Infinitesimal Isometries and Characteristic Numbers

Let M be an oriented Riemannian manifold of dimension $n=2m$. Let P be the bundle of oriented orthonormal frames over M; it is a principal bundle over M with group $\mathrm{SO}(n)$. Denote by Ω the curvature form on P. Let f be a symmetric form of degree p on the Lie algebra $\mathfrak{o}(n)$ which is $\mathrm{ad}(\mathrm{SO}(n))$-invariant. For the sake of simplicity, denote by $f(\Omega)$ the $2p$-form $f(\Omega, \dots, \Omega)$ on P. Then there exists a unique closed $2p$-form $\bar{f}(\Omega)$ on M such that $\pi^*(\bar{f}(\Omega))=f(\Omega)$, where $\pi\colon P \to M$ is the projection (see, for example, Kobayashi-Nomizu [1; Chapter XII]). The cohomology class defined by $\bar{f}(\Omega)$ is called the *characteristic class* defined by f. If $p=m$ and M is compact, then the integral of $\bar{f}(\Omega)$ over M is called the *characteristic number* defined by f.

Let X be an infinitesimal isometry of M. Let $\mathrm{Zero}(X)=\bigcup N_i$ be the zero set of X, where the N_i's are the connected components of $\mathrm{Zero}(X)$. As we are interested in one N_i for the moment, denote N_i by N. Let $2r$ be the codimension of N so that $\dim N = 2m-2r$. Let P_N be the bundle of adapted frames over N; it is a principal bundle over N with group $\mathrm{SO}(2m-2r)\times\mathrm{SO}(2r)$. (By an adapted frame, we mean an oriented orthonormal frame whose first $(2m-2r)$ basis elements are tangent to N and whose last $2r$ elements are normal to N.) The curvature form Ω restricted to the subbundle P_N of P is of the form

$$\begin{pmatrix} \Omega^a_b & 0 \\ 0 & \Omega^i_j \end{pmatrix} \qquad 1 \leqq a,\ b \leqq 2m-2r; \qquad 2m-2r+1 \leqq i,\ j \leqq 2m.$$

We recall that a tensorial p-form of type $\mathrm{ad}(\mathrm{SO}(n))$ is an $\mathfrak{o}(n)$-valued p-form φ on P such that

$$R_a^*(\varphi) = (\mathrm{ad}\, a^{-1})\,\varphi \qquad \text{for } a \in \mathrm{SO}(n),$$

where R_a denotes the right translation of P by $\mathrm{SO}(n)$ and

$$\varphi(Z_1, \dots, Z_p) = 0 \qquad \text{whenever } Z_1 \text{ is a vertical vector.}$$

For example, the curvature form Ω is a tensorial 2-form of type $\mathrm{ad}(\mathrm{SO}(n))$. Since the covariant derivative ∇X of X is a skew-symmetric linear endomorphism of the tangent bundle $T(M)$, it may be also considered as a tensorial 0-form of type $\mathrm{ad}(\mathrm{SO}(n))$ on P; this 0-form on P will be still denoted by ∇X. If we restrict the 0-form ∇X to P_N, then it is of the form

$$\nabla X = \begin{pmatrix} 0 & 0 \\ 0 & \Lambda \end{pmatrix},$$

where Λ is a matrix of order $2r$, see § 5.

Let f be an $\mathrm{ad}(SO(n))$-invariant symmetric form of degree p on the Lie algebra $\mathfrak{o}(n)$. Consider the form

$$f(t\Omega+\nabla X)$$

on P_N. If we denote by D the exterior covariant differentiation (Kobayashi-Nomizu [1, vol. 1; p. 77]), then not only $D\Omega=0$ but also $D(\nabla X)=0$ on P_N since ∇X is parallel on N as we have shown in Theorem 5.3. It follows that $f(t\Omega+\nabla X)$ is closed. Denote by $\bar{f}(t\Omega+\nabla X)$ the form on N defined by $\pi^*(\bar{f}(t\Omega+\nabla X))=f(t\Omega+\nabla X)$. We may write

$$\bar{f}(t\Omega+\nabla X)=\sum_k \binom{p}{k} \bar{f}(\underbrace{\Omega,\ldots,\Omega}_{k}, \underbrace{\nabla X,\ldots,\nabla X}_{p-k})\, t^k.$$

The coefficient of t^k is a closed $2k$-form on N; it is a polynomial in the curvature form with constant coefficients since ∇X is parallel on N.

Consider now the polynomial $\det(S)$, $S\in\mathfrak{o}(2p)$, on the Lie algebra $\mathfrak{o}(2p)$. We know that the determinant of a skew-symmetric matrix S is a square of a polynomial. More precisely, there is a unique polynomial $\sqrt{\det(S)}$ on $\mathfrak{o}(2p)$ such that $(\sqrt{\det(S)})^2=\det(S)$ and

$$\sqrt{\det(S)}=s_1 s_2 \ldots s_p \qquad \text{for } S=\begin{pmatrix} 0 & s_1 & & & \\ -s_1 & 0 & & & \\ & & \ddots & & \\ & & & 0 & s_p \\ & & & -s_p & 0 \end{pmatrix}.$$

In fact, (cf. Kobayashi-Nomizu [1, vol. 2; p. 304])

$$\sqrt{\det(S)}=\frac{1}{2^p\, p!}\sum \varepsilon_{i_1\ldots i_{2p}}\, s^{i_1}_{i_2}\ldots s^{i_{2p-1}}_{i_{2p}} \qquad \text{for } S=(s^i_j),$$

where $\varepsilon_{i_1\ldots i_{2p}}$ is the sign of the permutation $(1,\ldots,2p)\to(i_1,\ldots,i_{2p})$. We set

$$\chi_p(S)=\frac{1}{(2\pi)^p}(\sqrt{\det(S)}) \qquad \text{for } S\in\mathfrak{o}(2p).$$

We know that the curvature form Ω restricted to P_N splits into the tangential and normal parts. We denote by Ω_ν the normal part (Ω^i_j), $i,j=2m-2r+1,\ldots,2m$. We denote the normal part of ∇X by Λ as before. Since the form $\chi_r(t\Omega_\nu+\Lambda)$ on P_N is defined by an $\mathrm{ad}(SO(2r))$-invariant polynomial χ_r on $\mathfrak{o}(2r)$, there is a unique closed form $\bar{\chi}_r(t\Omega_\nu+\Lambda)$ on N such that $\pi^*(\bar{\chi}_r(t\Omega_\nu+\Lambda))=\chi_r(t\Omega_\nu+\Lambda)$. Since ∇X is parallel on N, $\bar{\chi}_r(t\Omega_\nu+\Lambda)$ expands into a polynomial in t whose coefficients are all polynomials in Ω^i_j, $i,j>2m-2r$, with constant coefficients. Since

$\det(\Lambda) \neq 0$, the constant term of $\bar{\chi}_r(t\Omega_\nu + \Lambda)$ is nonzero. Hence,

$$1/(\bar{\chi}_r(t\Omega_\nu + \Lambda))$$

can be expanded into a power series in t whose coefficients are forms on N which can be written as polynomials in Ω^i_j, $i, j > 2m - 2r$.

Let f be an ad(SO(2m))-invariant symmetric form of degree m on $\mathfrak{o}(2m)$. Then the residue $\mathrm{Res}_f(N)$ is defined by

$$\mathrm{Res}_f(N) \cdot t^{m-r} = \int_N \frac{\bar{f}(t\Omega + \nabla X)}{\bar{\chi}_r(t\Omega_\nu + \Lambda)}.$$

We shall now prove the following formula of Bott [1, 2]. See also Baum-Cheeger [1].

Theorem 6.1. *Let M be a compact, oriented Riemannian manifold of dimension $n = 2m$. Let X be an infinitesimal isometry of M with zero set* Zero$(X) = \bigcup N_i$, *where the N_i's are the connected components of* Zero(X). *Let f be an* (ad SO(n))*-invariant symmetric form of degree m on $\mathfrak{o}(n)$. Then the characteristic number $\int_M \bar{f}(\Omega)$ of M defined by f is given by*

$$\int_M \bar{f}(\Omega) = \sum_i \mathrm{Res}_f(N_i).$$

Proof. The first formula we are going to prove is the following:

$$t\,d\bar{f}(t\Omega + \nabla X) = \iota_X(\bar{f}(t\Omega + \nabla X)), \quad \text{on } M, \tag{1}$$

where ι_X is the interior product by X. We shall do all our calculations on the principal bundle P so that X and ∇X are also lifted to P in a natural manner. We denote by D the exterior covariant derivation. Then

$$\begin{aligned} t\,d(f(t\Omega + \nabla X)) &= t\,D(f(t\Omega + \nabla X)) = t\,D(f(t\Omega + \nabla X, \ldots, t\Omega + \nabla X)) \\ &= m\,t\,f(D\nabla X, t\Omega + \nabla X, \ldots, t\Omega + \nabla X) \\ &= m\,t\,f(\iota_X \Omega, t\Omega + \nabla X, \ldots, t\Omega + \nabla X) \\ &= m\,t\,f(\iota_X(t\Omega + \nabla X), t\Omega + \nabla X, \ldots, t\Omega + \nabla X) \\ &= \iota_X f(t\Omega + \nabla X). \end{aligned}$$

In the proof above of (1), we made use of the formula $D\nabla X = \iota_X \Omega$; this formula is equivalent to the formula in (1) of Proposition 2.2 but can be derived easily and directly from $L_X \omega = 0$ and $\iota_X \omega = \nabla X$, where ω is the connection form.

We define a 1-form ψ on M-Zero(X) by

$$\psi(Y) = g(X, Y)/g(X, X) \quad \text{for } Y \in T(M\text{-Zero}(X)).$$

Then

(2) $$\psi(X)=1, \quad L_X\psi=0 \quad \text{and} \quad \iota_X d\psi=0.$$

We set

$$\eta=\bar{f}(t\Omega+\nabla X)\frac{\psi}{1-t\,d\psi},$$

where $1/(1-t\,d\psi)$ means $1+t\,d\psi+t^2(d\psi)^2+\cdots$.

We prove the following formula:

(3) $$\bar{f}(t\Omega+\nabla X)+t\,d\eta-\iota_X\eta=0.$$

This is a consequence of (1) and the following two formulae:

$$t\,d\eta=t\,d\bar{f}(t\Omega+\nabla X)\frac{\psi}{1-t\,d\psi}+\bar{f}(t\Omega+\nabla X)\frac{t\,d\psi}{1-t\,d\psi}$$

and

$$\iota_X\eta=\iota_X\bar{f}(t\Omega+\nabla X)\frac{\psi}{1-t\,d\psi}+\bar{f}(t\Omega+\nabla X)\frac{1}{1-t\,d\psi},$$

which follows from (2).

We are now interested in the coefficient of t^m in the formula (3). We may write

$$\eta=\eta_{m-1}t^{m-1}+\eta_{m-2}t^{m-2}+\cdots,$$

where η_k is a $(2k+1)$-form. (Since $\dim M=2m$, the coefficient of t^k vanishes if k is greater than m). Then

$$t\,d\eta=d\eta_{m-1}t^m+\cdots,$$
$$\iota_X\eta=\iota_X\eta_{m-1}t^{m-1}+\cdots.$$

On the other hand, the coefficient of t^m in $\bar{f}(t\Omega+\nabla X)$ is given by $\bar{f}(\Omega)$. Hence, we have

(4) $$\bar{f}(\Omega)+d\eta_{m-1}=0 \quad \text{on } M\text{-Zero}(X).$$

Let $N_{i\varepsilon}$ denote the ε-neighborhood of N_i. Using (4) and Stokes formula, we obtain

$$\begin{aligned}\int_M\bar{f}(\Omega)&=\lim_{\varepsilon\to 0}\int_{M-\cup N_{i\varepsilon}}\bar{f}(\Omega)=-\lim_{\varepsilon\to 0}\int_{M-\cup N_{i\varepsilon}}d\eta_{m-1}\\&=\sum_i\lim_{\varepsilon\to 0}\int_{\partial N_{i\varepsilon}}\eta_{m-1}.\end{aligned}$$

To complete the proof of the theorem, we have to show that

$$\operatorname{Res}_f(N_i)=\lim_{\varepsilon\to 0}\int_{\partial N_{i\varepsilon}}\eta_{m-1}.$$

Since we are now interested in each individual N_i, we denote N_i by N. The boundary ∂N_ε of the ε-neighborhood of N is a sphere bundle over N with fibre S^{2r-1}, where $2r$ is the codimension of N. Let

$$\sigma_\varepsilon\colon \partial N_\varepsilon \to N$$

be the projection. If we denote by $\Phi(\partial N_\varepsilon)$ and $\Phi(N)$ the algebras of differential forms on ∂N_ε and N, respectively, then we have a natural algebra homomorphism

$$\sigma_\varepsilon^*\colon \Phi(N) \to \Phi(\partial N_\varepsilon).$$

On the other hand, the integration over the fibre in $\partial N_\varepsilon \to N$ will be denoted by σ_*^ε. Hence

$$\sigma_*^\varepsilon\colon \Phi(\partial N_\varepsilon) \to \Phi(N)$$

is a linear mapping which decreases degrees by $2r-1$. Moreover, σ_*^ε is characterized by the following formula:

$$\int_{\partial N_\varepsilon} u \cdot \sigma_\varepsilon^* v = \int_N \sigma_*^\varepsilon u \cdot v \qquad \text{for } u \in \Phi(\partial N_\varepsilon),\ v \in \Phi(N).$$

If φ is a form defined on M-Zero(X), then we denote by $\sigma_*^\varepsilon(\varphi)$ the integral over the fibre of the restriction of φ to ∂N_ε. We set

$$\sigma_*(\varphi) = \lim_{\varepsilon \to 0} \sigma_*^\varepsilon(\varphi)$$

provided the limit exists.

To calculate the integral

$$\lim_{\varepsilon \to 0} \int_{\partial N_\varepsilon} \eta_{m-1},$$

we first integrate η_{m-1} over the fibre in $\partial N_\varepsilon \to N$ and then integrate the result over the base N. Hence, we are interested in $\sigma_*(\eta_{m-1})$. But this is the coefficient of t^{m-1} in $\sigma_* \eta$. We are now interested in calculating

$$\sigma_* \eta = \sigma_* \left(\bar{f}(t\Omega + \nabla X) \cdot \frac{\psi}{1 - t\,d\psi} \right).$$

Since $\bar{f}(t\Omega + \nabla X)$ is smoothly defined on the entire space M including N, it follows that

$$\text{(5)} \qquad \sigma_* \eta = \bar{f}(t\Omega + \nabla X)|_N \cdot \sigma_* \left(\frac{\psi}{1 - t\,d\psi} \right).$$

The problem now is to evaluate $\sigma_*(\psi/1 - t\,d\psi)$. Since

$$\frac{\psi}{1 - t\,d\psi} = \psi + t\psi \wedge d\psi + t^2 \psi \wedge (d\psi)^2 + \cdots,$$

the problem is further reduced to that of calculating $\sigma_*(\psi \wedge (d\psi)^{q-1})$.

Let ξ be the 1-form corresponding to the infinitesimal isometry X under the duality defined by the metric tensor. Then

$$\psi = \frac{\xi}{\|\xi\|^2}$$

and

$$d\psi = \frac{\|\xi\|^2 \, d\xi - d(\|\xi\|^2) \wedge \xi}{\|\xi\|^4}$$

so that

(6) $$\psi \wedge (d\psi)^{q-1} = \frac{\xi \wedge (d\xi)^{q-1}}{\|\xi\|^{2q}}.$$

We fix a point o of N and let $x^1, \ldots, x^{2p}$, $y^1, \ldots, y^{2r}$, $m = p + r$, be a normal coordinate system around o such that N is locally defined by $y^1 = \cdots = y^{2r} = 0$. Such a normal coordinate system exists since N is totally geodesic. We consider the Taylor expansions of ξ and $d\xi$ at o. Then

(7) $$\xi = \sum \Lambda_{ij} \, y^i \, dy^j + \cdots,$$

(8) $$d\xi = \sum \Lambda_{ij} \, dy^i \wedge dy^j - \sum R^i_{jab} \Lambda_{ik} \, y^k \, y^j \, dx^a \wedge dx^b + \cdots,$$

where the dots indicate terms with total degree in y and dy greater than 2 and (Λ_{ij}) is a skew-symmetric non-degenerate matrix. These two formulae may be proved as follows. If we denote $y^1, \ldots, y^{2r}$ by $x^{2p+1}, \ldots, x^{2p+2r}$ and write $\xi = \sum \xi_A \, dx^A$, then

$$(\xi_A)_o = 0,$$

$$\xi_{A;B} + \xi_{B;A} = 0 \qquad \text{(2) of Proposition 2.2,}$$

$$\xi_{A;B;C} + \sum R^D_{CAB} \, \xi_D = 0 \qquad \text{(1) of Proposition 2.2,}$$

$$(\xi_{A;B})_o = 0 \quad \text{unless } A, B \geqq 2p+1 \qquad \text{(Theorem 5.3),}$$

$$(\xi_{A;B})_{A,B=2p+1,\ldots,2p+2r} \quad \text{is non-degenerate at o} \qquad \text{(Theorem 5.3),}$$

$$\left(\sum_{a,b=1}^{2p} R^D_{Cab} \, dx^a \wedge dx^b \right)_o = 0 \quad \text{unless } C, D \leqq 2p$$

$$\text{or} \quad C, D \geqq 2p+1 \qquad (N \text{ is totally geodesic}).$$

From these formulae, both (7) and (8) can be easily obtained. It is clear that in calculating $\sigma_*(\psi \wedge (d\psi)^{q-1})$ we can replace ξ and $d\xi$ by their Taylor expansions and ignore the terms of degree sufficiently high in y and dy. It is indeed not difficult to see that the terms indicated by dots in (7) and (8) can be ignored. (For the detail on this point, see Bott [2; pp. 321–323].)

We set

$$\alpha=\sum \Lambda_{ij}\, y^i\, d y^j,$$
$$\beta=\sum \Lambda_{ij}\, d y^i \wedge d y^j,$$
$$\gamma=\sum R^i_{jab}\, \Lambda_{ik}\, y^k\, y^j\, d x^a \wedge d x^b$$

so that ξ and $d\xi$ are approximated by α and $\beta-\gamma$, respectively. From (6), we obtain

$$\sigma_*(\psi\wedge(d\psi)^{q-1})=\sigma_*\left(\frac{\alpha\wedge(\beta-\gamma)^{q-1}}{\|\alpha\|^{2q}}\right). \tag{9}$$

Since the dimension of the fibre in $\partial N_\varepsilon\to N$ is $2r-1$, the integration over the fibre annihilates any term whose degree in dy is not exactly $2r-1$. This reduces (9) to

$$\sigma_*(\psi\wedge(d\psi)^{q-1})=0 \qquad \text{for } q<r,$$
$$\sigma_*(\psi\wedge(d\psi)^{q-1})=\binom{q-1}{q-r}\cdot\sigma_*\left(\frac{\alpha\wedge\beta^{r-1}\wedge(-\gamma)^{q-r}}{\|\alpha\|^{2q}}\right) \qquad \text{for } q\geqq r. \tag{10}$$

Hence,

$$\sigma_*\left(\frac{\psi}{1-t\,d\psi}\right)=\sigma_*\left(\sum t^{q-1}\,\psi\wedge(d\psi)^{q-1}\right)$$
$$=\sigma_*\left(\sum_{q=r}^{\infty}\binom{q-1}{q-r}\frac{\alpha\wedge\beta^{r-1}\wedge(-\gamma)^{q-r}}{\|\alpha\|^{2q}}\,t^{q-1}\right) \tag{11}$$
$$=\sigma_*\left(\frac{\alpha\wedge\beta^{r-1}}{(\|\alpha\|^2+\gamma t)^r}\,t^{r-1}\right).$$

We set

$$z_i=\sum \Lambda_{ij}\, y^j,$$
$$C^{ij}=\sum R^i_{kab}\,\Lambda^{kj}\, d x^a\wedge d x^b,$$

where (Λ^{ij}) is the inverse matrix of (Λ_{ij}). (In the sequel, by fixing dx^a we consider (C^{ij}) as a constant matrix rather than a matrix valued 2-form at o. In other words, C^{ij} means actually $C^{ij}(U,V)$ where U and V are arbitrarily chosen tangent vectors of N at o.) We claim that (C^{ij}) is symmetric, i.e., $C^{ij}=C^{ji}$. In fact, since the 1-parameter group $\exp((t\nabla X)_o)$ of linear transformations of $T_o(M)$ preserves the curvature tensor at o, we have

$$\sum R^i_{kab}\,\Lambda^{kj}=\sum \Lambda^{ik}\, R_{kjab} \qquad \text{for } a,b\leqq 2m-2r \quad \text{and} \quad i,j,k>2m-2r.$$

Our assertion follows from the fact that $\Lambda^{ik}=-\Lambda^{ki}$ and $R_{kjab}=-R_{jkab}$. Now we can write $\|\alpha\|^2$ and γ as follows:

$$\alpha^2=\|z\|^2 \qquad (=\textstyle\sum z_i z_i),$$
$$\gamma=C(z,z) \quad (=\textstyle\sum C^{ij} z_i z_j).$$

Choosing $y^1, \ldots, y^{2r}$ in such a way that $\Lambda_{ij}=0$ except for $i=2s-1, j=2s$ or $i=2s, j=2s-1$ where $s=1, \ldots, r$, we can easily verify the formula:

$$\alpha \wedge \beta^{r-1} = -(r-1)! \frac{2^{r-1}}{\sqrt{\det(\Lambda)}} v(z),$$

where

$$\Lambda=(\Lambda_{ij}), \quad v(z)=\sum_i (-1)^i \, dz_1 \wedge \cdots \wedge dz_{i-1} \wedge z_i \, dz_{i+1} \wedge \cdots \wedge dz_{2r}.$$

Then

$$\frac{\alpha \wedge \beta^{r-1}}{(\|\alpha\|^2+\gamma t)^2} = -(r-1)! \, 2^{r-1} \frac{\varphi(z;t)}{\sqrt{\det(\Lambda)}}, \tag{12}$$

where

$$\varphi(z;t)=\frac{v(z)}{\|z\|^2+C(z,z)t}.$$

For each fixed t, $\varphi(z;t)$ is a $(2r-1)$-form which is defined and closed on $\mathbf{R}^{2r}-0$. It follows that, for any closed hypersurface H in $\mathbf{R}^{2r}-0$ homotopic to the unit sphere $\|z\|=1$,

$$\int_H \varphi(z;t) = \int_{\|z\|=1} \varphi(z;t). \tag{13}$$

We set

$$h(t)=\int_{\|z\|=1} \varphi(z;t).$$

If t is fixed to be a sufficiently small constant, then the quadratic form $\|z\|^2+C(z,z)t$ is positive definite. Hence, there exists a linear automorphism A_t of $\mathbf{R}^{2r}$ such that ${}^tA_t A_t = I + Ct$, where I denotes the identity matrix of order $2r$. We may impose even the condition that A_t be orientation-preserving, i.e., $\det(A_t)>0$. If we set

$$w=A_t z,$$

then

$$v(w)=\det(A_t)\cdot v(z),$$

$$\|z\|^2+C(z,z)t=\|w\|^2$$

so that

$$\int_{\|A_t z\|=1} \frac{v(z)}{(\|z\|^2+C(z,z)t)^r} = \int_{\|w\|=1} \frac{v(w)}{\det(A_t)\cdot\|w\|^{2r}}.$$

But the left hand side is equal to $h(t)$ by (13). Hence,

$$h(t)=\frac{1}{\det(A_t)} \int_{\|w\|=1} v(w). \tag{14}$$

From the definition of A_t, we obtain

$$\det(A_t)^2 = \det(I + Ct).$$

We denote (Λ_{ij}) by Λ and $(\sum R^i_{jab}\, dx^a \wedge dx^b)$ by Ω_v so that $(\Lambda^{ij}) = \Lambda^{-1}$ and $(C^{ij}) = \Omega_v \cdot \Lambda^{-1}$. Then

$$\begin{aligned}\det(I + Ct) &= \det(\Lambda^{-1}) \det(\Lambda + \Omega_v t)\\ &= (\chi_r(\Lambda)^{-1} \chi_r(\Lambda + \Omega_v t))^2.\end{aligned}$$

Since $\det(A_t)$ is positive, we can conclude that

$$\det(A_t) = \chi_r(\Lambda)^{-1} \chi_r(\Lambda + \Omega_v t). \tag{15}$$

By Stokes formula, we obtain

$$\begin{aligned}\int_{\|w\|=1} v(w) &= \int_{\|w\| \leq 1} d(v(w))\\ &= \int_{\|w\| \leq 1} (-2r)\, dw_1 \wedge dw_2 \wedge \cdots \wedge dw_{2r} = (-2r)\frac{\pi^r}{r!}.\end{aligned} \tag{16}$$

From (13) and the definition of $h(t)$, we have

$$\sigma_*(\varphi(z;t)) = \lim_{\varepsilon \to 0} \int_{\|z\|=\varepsilon} \varphi(z;t) = \int_{\|z\|=1} \varphi(z;t) = h(t). \tag{17}$$

From (11), (12) and (17), we obtain

$$\sigma_*\left(\frac{\psi}{1 - t\,d\psi}\right) = -t^{r-1} \frac{(r-1)!\, 2^{r-1}}{\sqrt{\det(\Lambda)}} h(t). \tag{18}$$

From (14), (15), (16) and (18), we obtain

$$\sigma_*\left(\frac{\psi}{1 - t\,d\psi}\right) = \frac{1}{\bar{\chi}_r(\Lambda + t\Omega_v)} t^{r-1}. \tag{19}$$

From (5) and (19), we obtain

$$\sigma_* \eta = \frac{\bar{f}(t\Omega + \nabla X)|_N}{\bar{\chi}_r(t\Omega_v + \Lambda)} t^{r-1}. \tag{20}$$

Since η_{m-1} is the coefficient of t^{m-1} in η, it follows from (20) that $\sigma_*(\eta^{m-1})$ is the coefficient of t^{m-r} in

$$\frac{\bar{f}(t\Omega + \nabla X)|_N}{\bar{\chi}_r(t\Omega_v + \Lambda)}.$$

Hence,

$$\lim_{\varepsilon \to 0} \int_{\partial N_\varepsilon} \eta_{m-1}\, t^{m-r} = \int_N \sigma_*(\eta_{m-1})\, t^{m-r} = \int_N \frac{\bar{f}(t\Omega + \nabla X)}{\chi_r(t\Omega_v + \Lambda)}.$$

But the right hand side is equal to $\mathrm{Res}_f(N)\,t^{m-r}$ by definition. Hence,

$$\lim_{\varepsilon\to 0}\int_{\partial N_\varepsilon}\eta_{m-1}=\mathrm{Res}_f(N).$$

This completes the proof. q.e.d.

If N is an isolated point $\{p\}$, then

$$\mathrm{Res}_f(p)=\frac{\bar{f}(\nabla X)}{\bar{\chi}_m(\Lambda)}.$$

This means that if $\mathrm{Zero}(X)$ consists of isolated points only, then the characteristic numbers of M can be expressed in terms of ∇X $(=\Lambda)$ at these isolated zero points. Even when an invariant polynomial f is so chosen that $\bar{f}(\Omega)$ represents an integral class, the individual $\mathrm{Res}_f(p)$ need not be an integer. It is difficult to compute $\mathrm{Res}_f(p)$ unless M is a symmetric space. If $\mathrm{Zero}(X)$ is empty, then the right hand side of Theorem 6.1 vanishes. Hence,

Corollary 6.2. *If a compact, orientable Riemannian manifold admits an infinitesimal isometry with empty zero set, then its Pontrjagin numbers vanish.*

The proof of Theorem 6.1 given here yields the following result. Let G be a Lie subgroup of $SO(2m)$ and P be a G-structure on a $2m$-dimensional manifold M. Assume that there is a torsionfree connection in P. Then Theorem 6.1 is valid for an infinitesimal automorphism X of the G-structure P and for an $\mathrm{ad}(G)$-invariant polynomial f on $\mathfrak{g}$ of degree m. Note that, since $G\subset SO(2m)$, an infinitesimal automorphism X of P is an infinitesimal isometry but that an $\mathrm{ad}(G)$-invariant polynomial on $\mathfrak{g}$ may not be induced from an $\mathrm{ad}(SO(2m))$-invariant polynomial on $\mathfrak{so}(2m)$. In particular, Theorem 6.1 is valid when $G=U(m)$ and P is a Kähler structure on M.

For a completely different proof of Theorem 6.1, see Atiyah-Singer [1].

III. Automorphisms of Complex Manifolds

1. The Group of Automorphisms of a Complex Manifold

Let M be a complex manifold and $\mathfrak{H}(M)$ the group of holomorphic transformations of M. In general, $\mathfrak{H}(M)$ can be infinite dimensional. For instance, $\mathfrak{H}(\mathbf{C}^n)$ is not a Lie group if $n \geqq 2$. To see this, consider transformations of $\mathbf{C}^2$ of the form

$$\begin{aligned} z' &= z \\ w' &= w + f(z) \end{aligned} \qquad (z, w) \in \mathbf{C}^2,$$

where $f(z)$ is an entire function in z, e.g., a polynomial of any degree in z. The fact that $\mathfrak{H}(\mathbf{C}^2)$ contains these transformations shows that $\mathfrak{H}(\mathbf{C}^2)$ cannot be finite dimensional. Similarly, for $\mathfrak{H}(\mathbf{C}^n)$ with $n \geqq 2$. On the other hand, $\mathfrak{H}(\mathbf{C})$ is the group of orientation preserving conformal transformations and, as we shall see later, it is a Lie group. The purpose of this section is to give conditions on M which imply that $\mathfrak{H}(M)$ is a Lie group.

Theorem 1.1. *Let M be a compact complex manifold. Then the group $\mathfrak{H}(M)$ of holomorphic transformations of M is a complex Lie transformation group and its Lie algebra consists of holomorphic vector fields on M.*

Proof. From Corollary 4.2 of Chapter I, we know that $\mathfrak{H}(M)$ is a Lie transformation group. Its Lie algebra can be identified with the Lie algebra of holomorphic vector fields; if $Z = X + iX$ is a holomorphic vector field with X and Y real, then X is an infinitesimal automorphism of the complex structure of M, and vice versa. So the Lie algebra of $\mathfrak{H}(M)$ is a complex Lie algebra; if Z is a holomorphic vector field, so is iZ. In other words, if X is an infinitesimal automorphism of the complex structure, so is JX. Hence, $\mathfrak{H}(M)$ is a complex Lie transformation group. q.e.d.

Theorem 1.1 is due to Bochner-Montgomery [2, 3]; they have actually shown that the topology of $\mathfrak{H}(M)$ is the compact-open topology.

We consider now the following theorem of H. Cartan [1, 2]:

Theorem 1.2. *Let M be a bounded domain in $\mathbf{C}^n$. Then the group $\mathfrak{H}(M)$ of holomorphic transformations of M is a Lie transformation group and the isotropy subgroup $\mathfrak{H}_x(M)$ of $\mathfrak{H}(M)$ at any point $x \in M$ is compact. If X is in the Lie algebra of $\mathfrak{H}(M)$, then JX is not in the Lie algebra of $\mathfrak{H}(M)$.*

We shall elaborate a little on the last statement. If X is an infinitesimal automorphism, then JX is also an infinitesimal automorphism. In other words, if Z is a holomorphic vector field, then iZ is obviously also holomorphic. In order that X belongs to the Lie algebra of $\mathfrak{H}(M)$, X must be complete, i.e., must generate a *global* 1-parameter group of *global* holomorphic transformations. The last statement in Theorem 1.2 means that if X is complete, then JX cannot be complete. This should be viewed in contrast to Theorem 1.1, where M is compact so that every vector field is complete.

We shall now give two results each of which generalizes Theorem 1.2. We recall first the definition of Bergman metric. Let M be an n-dimensional complex manifold and H the complex Hilbert space of holomorphic n-forms f which are square integrable in the sense that

$$\int_M i^{n^2} f \wedge \bar{f} < \infty .$$

The inner product of H is given by

$$(f, g) = \int_M i^{n^2} f \wedge \bar{g} .$$

We assume that H is *very ample* in the following sense:

(1) At each point x of M, there exists an $f \in H$ such that $f(x) \neq 0$.

(2) If $z^1, \ldots, z^n$ is a local coordinate system in a neighborhood of a point $x \in M$, then, for each j, there exists an element

$$h = h^* \, dz^1 \wedge \cdots \wedge dz^n$$

of H such that $h(x) = 0$ and $(\partial h^* / \partial z^j)_x \neq 0$.

Let $h_0, h_1, h_2, \ldots$ be a complete orthonormal basis for the Hilbert space H and define the *Bergman kernel form* K by

$$K = K^* \, dz^1 \wedge \cdots \wedge dz^n \wedge d\bar{z}^1 \wedge \cdots \wedge d\bar{z}^n = \sum_{k=0}^{\infty} h_k \wedge \bar{h}_k .$$

(If M is a domain in $\mathbf{C}^n$ with natural coordinate system $z^1, \ldots, z^n$, the function K^* is the classical *Bergman kernel function* of M.) We define the Bergman metric ds^2 of M by

$$ds^2 = 2 \sum_{\alpha, \beta = 1}^{n} g_{\alpha\bar{\beta}} \, dz^\alpha \cdot d\bar{z}^\beta, \qquad \text{where } g_{\alpha\bar{\beta}} = \partial^2 \log K^* / \partial z^\alpha \partial \bar{z}^\beta .$$

It is not hard to see that K is defined independent of the choice of $h_0, h_1, h_2, \ldots$ and the metric ds^2 is independent of the choice of $z^1, \ldots, z^n$. We remark that Condition (1) guarantees that $K \neq 0$ everywhere, i.e., $K^* > 0$ everywhere so that ds^2 is defined and Condition (2) implies that ds^2 is positive definite. (Without (2), ds^2 is, in general, positive semi-definite.) A more geometric interpretation can be given to (1) and (2) as follows. For each point x of M, let $H(x)$ denote the subspace of H consisting of those holomorphic forms f vanishing at x. Condition (1) says that $H(x)$ is a hyperplane in H. The set of all hyperplanes in H forms a complex projective space (possibly of infinite dimension). Since this projective space is isomorphic, in a natural manner, to the projective space of complex lines in the dual space H^* of H, we denote it by $P(H^*)$. Then we have a mapping $M \to P(H^*)$ which sends x into $H(x)$. From the definition of the Fubini-Study metric, it follows that ds^2 is induced from the Fubini-Study metric of $P(H^*)$ by the mapping $M \to P(H^*)$. Condition (2) says that the mapping $M \to P(H^*)$ is an immersion. For more details, see Kobayashi [6].

Theorem 1.3. *Let M be a complex manifold of dimension n such that the space of square-integrable holomorphic n-forms is very ample (so that the Bergman metric is defined). Then the group $\mathfrak{H}(M)$ of holomorphic transformations of M is a Lie transformation group and the isotropy subgroup $\mathfrak{H}_x(M)$ of $\mathfrak{H}(M)$ at any point $x \in M$ is compact. If X is a nonzero element of the Lie algebra of $\mathfrak{H}(M)$, then JX is not in the Lie algebra of $\mathfrak{H}(M)$ provided one of the following three conditions is satisfied:*

(a) *There is no parallel vector field (with respect to the Bergman metric) on M.*

(b) *There is no holomorphic mapping of $\mathbf{C}$ into M except the constant mappings.*

(c) *There is a point of M where the Ricci tensor is non-degenerate.*

Proof. Clearly, $\mathfrak{H}(M)$ is a closed subgroup of the group $\mathfrak{I}(M)$ of isometries of M with respect to the Bergman metric. It follows from Theorem 1.2 of Chapter II that $\mathfrak{H}(M)$ is a Lie transformation group and $\mathfrak{H}_x(M)$ is compact.

Lemma. *If X and JX are infinitesimal automorphisms of a Kähler manifold M, then X is parallel.*

Proof of Lemma. We recall the definition of A_X (cf. § 2 of Chapter II):

$$A_X = L_X - \nabla_X .$$

Then

$$J \circ A_X = A_X \circ J = A_{JX},$$

where the first equality follows from the definition of A_X and the second equality is a consequence of the formula $A_X Y = -\nabla_Y X$ in Proposition 2.1 of Chapter II. By Proposition 2.2 of Chapter II, we have the equation

$$g(A_{JX} JY, Z) + g(JY, A_{JX} Z) = 0 \quad \text{for all vector fields } Y, Z,$$

which can be easily transformed into

$$-g(A_X Y, Z) + g(Y, A_X Z) = 0.$$

On the other hand, from Proposition 2.1 of Chapter II we have

$$g(A_X Y, Z) + g(Y, A_X Z) = 0.$$

Hence, $g(A_X Y, Z) = 0$ for all Y, Z, and, consequently, $A_X = 0$. Since $A_X Y = -\nabla_Y X$, this shows that X is parallel, thus completing the proof of Lemma.

To complete the proof of Theorem 1.3, assume that the Lie algebra of $\mathfrak{H}(M)$ contains both X and JX. From Lemma it follows that if (a) is satisfied, then $X = 0$. Suppose (b) is satisfied. Since X and JX commute and both generate global 1-parameter group of holomorphic transformations of M, they generate a 1-dimensional complex Lie group acting holomorphically on M. Since every 1-dimensional complex Lie group has $\mathbf{C}$ as the universal covering group, we obtain a holomorphic action of $\mathbf{C}$ on M. By (b), this action is trivial and, hence, $X = 0$. Suppose (c) is satisfied. If $X \neq 0$, then X is parallel and M has a flat factor in its de Rham decomposition (locally). This would imply that the Ricci tensor is degenerate. q.e.d.

Remark. It is clear that if M is a bounded domain in $\mathbf{C}^n$, then (b) is satisfied. It is not known if (a), (b) or (c) can be removed in the theorem above. We mention two important cases where (c) is satisfied:

(1) M is homogeneous, i.e., $\mathfrak{H}(M)$ is transitive on M.

(2) There is a discrete subgroup Γ of $\mathfrak{H}(M)$ acting freely on M such that the quotient manifold M/Γ is compact. (In particular, the case where M is compact is contained in this case.)

To see that (c) is satisfied in the two cases above, let

$$V = V^* \, dz^1 \wedge \cdots \wedge dz^n \wedge d\bar{z}^1 \wedge \cdots \wedge d\bar{z}^n$$

be the volume element on M defined by the Bergman metric ds^2. We recall that the components of the Ricci tensor are given by

$$R_{\alpha\bar{\beta}} = -\partial^2 \log V^* / \partial z^\alpha \partial \bar{z}^\beta.$$

We compare this with the definition of the components $g_{\alpha\bar{\beta}}$ of the Bergman metric. If M is homogeneous, an invariant volume element is unique up to a constant factor so that $V^* = c\,K^*$, where c is a nonzero constant. Hence, $R_{\alpha\bar\beta} = -g_{\alpha\bar\beta}$. In case (2), K and V can be considered as $2n$-forms on M/Γ since they are invariant by Γ. Then the 2-forms

$$\frac{1}{2\pi i}\sum g_{\alpha\bar\beta}\, dz^\alpha \wedge d\bar z^\beta \quad \text{and} \quad -\frac{1}{2\pi i}\sum R_{\alpha\bar\beta}\, dz^\alpha \wedge d\bar z^\beta$$

on M/Γ define the same cohomology class, the first Chern class $c_1(M/\Gamma)$, of M/Γ. If we use $g_{\alpha\bar\beta}$, we see immediately that $c_1(M/\Gamma)^n \neq 0$. On the other hand, if $\det(R_{\alpha\bar\beta}) = 0$ everywhere, then $c_1(M/\Gamma)^n = 0$, which is a contradiction.

To state another generalization of the theorem of H. Cartan, we define a certain intrinsic pseudo-distance on a complex manifold M. Let D be an open unit disk in $\mathbf{C}$ with Poincaré distance (i.e., non-Euclidean distance) ρ. Given two point p and q of M, we choose a sequence of points $p = p_0, p_1, \ldots, p_{k-1}, p_k = q$ in M, points $a_1, \ldots, a_k, b_1, \ldots, b_k$ in the disk D and holomorphic mappings $f_1, \ldots, f_k$ of D into M such that $f_i(a_i) = f_{i-1}(b_{i-1}) = p_{i=1}$ for $i = 1, 2, \ldots, k$ and $f_k(b_k) = p_k$. We set

$$d_M(p, q) = \inf\left(\sum_{i=1}^{k} \rho(a_i, b_i)\right),$$

where the infimum is taken with respect to all possible choices for p_i, a_i, b_i, f_i above. Then d_M is a pseudo-distance; it is symmetric and satisfies the triangular axiom. For details on this pseudo-distance, we refer the reader to Kobayashi [9, 10]. If d_M is a distance, then M is called a *hyperbolic manifold.* A hyperbolic manifold M is said to be *complete* if d_M is a complete distance.

Theorem 1.4. *Let M be a hyperbolic manifold. Then the group $\mathfrak{H}(M)$ of holomorphic transformations of M is a Lie transformation group and the isotropy subgroup $\mathfrak{H}_x(M)$ of $\mathfrak{H}(M)$ at any point $x \in M$ is compact. If X is a nonzero element of the Lie algebra of $\mathfrak{H}(M)$, then JX is not in the Lie algebra of $\mathfrak{H}(M)$.*

Proof. We make use of the following basic property of d_M.

Lemma. *If M and N are complex manifolds and f: $M \to N$ is a holomorphic mapping, then*

$$d_N(f(p), f(q)) \leqq d_M(p, q) \quad \textit{for } p, q \in M.$$

This lemma follows immediately from the definition of d_M and d_N.

In particular, if $M = N$ and f: $M \to M$ is a biholomorphic mapping, then f is an isometry with respect to d_M. Hence, $\mathfrak{H}(M)$ is a closed subgroup of the group $\mathfrak{I}(M)$ of isometries, which is known to be a Lie group by Theorem 3.3 of Chapter I and Theorem 1.1 of Chapter II. It follows that

$\mathfrak{H}(M)$ is also a Lie group. By Theorem 1.1 of Chapter II, $\mathfrak{H}_x(M)$ is compact. Assume X is a nonzero element of the Lie algebra of $\mathfrak{H}(M)$ such that JX is also in the Lie algebra of $\mathfrak{H}(M)$. Then X and JX generate a 1-dimensional complex Lie group. Taking its universal covering group, we may assume that the group $\mathbf{C}$ acts on M. For each point p of M, we consider the orbit of the group $\mathbf{C}$ through p. In this way, we obtain a holomorphic mapping of $\mathbf{C}$ into M. But the pseudo-distance $d_{\mathbf{C}}$ on $\mathbf{C}$ is trivial, i.e., $d_{\mathbf{C}}=0$. It follows from Lemma that every holomorphic mapping of $\mathbf{C}$ into M is a constant mapping. Hence, the orbit of $\mathbf{C}$ through p reduces to the single point p. This means that $\mathbf{C}$ acts trivially on M, i.e., X is the zero vector field on M. q.e.d.

From the differential geometric standpoint, the most interesting example of hyperbolic manifold is given by a hermitian manifold with holomorphic sectional curvature bounded above by a negative constant (Kobayashi [10]). For results essentially equivalent to Theorem 1.4, see also Kaup [4], Wu [1]. For more details on holomorphic transformations of bounded domains, see Kaup [1].

For a generalization of Theorem 1.1 to compact complex spaces (with singularities), see Gunning [1], Kerner [1]. For generalizations of Theorem 1.2 and 1.4 to complex spaces, see Kaup [1, 4]. H. Fujimoto [1] unifies all these generalizations. For automorphisms of special domains (homogeneous bounded domains, Siegel domains), see Pyatetzki-Shapiro [1], Kaup-Matsushima-Ochiai [1], Kaneyuki [1], Tanaka [8].

In connection with Theorem 1.2, for the case where the Bergman kernel form is positive but the Bergman metric is only semi-positive, see Lichnerowicz [5]. For automorphisms of a complex manifold with volume element, see Koszul [1].

2. Compact Complex Manifolds with Finite Automorphism Groups

It has been known for a long time that the automorphism group of a compact Riemann surface of genus greater than 1 is a finite group, Klein (see Poincaré [1]), Hurwitz [1]. In this section, we shall generalize this classical result to higher dimensional compact complex manifolds. The first generalization is the following (Kobayashi [7]).

Theorem 2.1. *Let M be a compact complex manifold with negative first Chern class. Then the group $\mathfrak{H}(M)$ of holomorphic transformations of M is finite.*

Before we proceed with the proof of the theorem, we shall elaborate on the assumption of "*negative first Chern class*". We say that the first Chern class $c_1(M)$ of M is negative if it can be represented by a closed (1, 1)-form

$$\frac{i}{2\pi}\sum \gamma_{\alpha\beta}\, dz^\alpha \wedge d\bar{z}^\beta$$

such that $(\gamma_{\alpha\beta})$ is everywhere negative definite. If $2\sum g_{\alpha\beta}\, dz^\alpha d\bar{z}^\beta$ is a hermitian metric on M, then $c_1(M)$ can be represented by

$$\frac{i}{2\pi}\sum R_{\alpha\beta}\, dz^\alpha \wedge d\bar{z}^\beta, \qquad \text{where } R_{\alpha\bar{\beta}} = -\partial^2 \log(\det(g_{\gamma\bar{\delta}}))/\partial z^\alpha \partial \bar{z}^\beta.$$

This shows that a hermitian manifold with negative definite Ricci tensor has negative first Chern class. The $2n$-form

$$i^{n^2} G\, dz^1 \wedge \cdots \wedge dz^n \wedge d\bar{z}^1 \wedge \cdots \wedge d\bar{z}^n, \qquad G = \det(g_{\alpha\beta}),$$

is the volume element of the hermitian metric $2\sum g_{\alpha\beta}\, dz^\alpha d\bar{z}^\beta$. More generally, if

$$i^{n^2}\, V dz^1 \wedge \cdots \wedge dz^n \wedge d\bar{z}^1 \wedge \cdots \wedge d\bar{z}^n$$

is any volume element of a compact complex manifold M, i.e., if $V>0$, then the 2-form

$$\frac{i}{2\pi}\sum \gamma_{\alpha\beta}\, dz^\alpha \wedge d\bar{z}^\beta, \qquad \text{where } \gamma_{\alpha\beta} = -\partial^2 \log V/\partial z^\alpha \partial \bar{z}^\beta,$$

represents $c_1(M)$.

In order to prove Theorem 2.1, we have to reformulate the assumption of "negative first Chern class" in algebraic terms. Let K denote the canonical line bundle of a compact complex manifold M; by definition, a local holomorphic section of K is a locally defined holomorphic n-form (where $n = \dim M$). The line bundle K is said to be *ample* if there exists a positive integer p such that the line bundle $K^p = K \otimes \cdots \otimes K$ is very ample in the following sense. Let H be the space of holomorphic sections of K^p; $H = H^0(M; K^p)$. At each point x of M, consider the subspace $H(x)$ of H consisting of sections vanishing at x. Then the condition is that $H(x)$ is a hyperplane of H for each x and the mapping $x \to H(x)$ gives an imbedding of M into the complex projective space $P(H^*)$ of hyperplanes in H. (Since the hyperplanes in H are in a natural one-to-one correspondence with the complex lines in the dual space H^* of H, the notation $P(H^*)$ is justified.) We recall that the Bergman metric exists on M if K itself is very ample (see § 1).

We claim that *the canonical line bundle K is ample if and only if $c_1(M)$ is negative.* The implication "K ample $\to c_1(M) < 0$" is trivial. Let $\omega_0, \ldots, \omega_N$ be a basis for H. If we write formally

$$\omega_0 \bar{\omega}_0 + \cdots + \omega_N \bar{\omega}_N = V \cdot (dz^1 \wedge \cdots \wedge dz^n)^p \cdot (d\bar{z}^1 \wedge \cdots \wedge d\bar{z}^n)^p,$$

then the 2-form

$$\frac{1}{2\pi i}\sum \gamma_{\alpha\bar\beta}\, dz^\alpha \wedge d\bar z^\beta, \qquad \text{where } \gamma_{\alpha\bar\beta} = -\partial^2 \log V/\partial z^\alpha \partial \bar z^\beta$$

represents the characteristic class of K^p, which is equal to $-p \cdot c_1(M)$. A simple local calculation shows that if K^p is very ample, then $(\gamma_{\alpha\bar\beta})$ is negative definite. The implication "$c_1(M)<0 \to K$ ample" is a result of Kodaira [2] and will not be proved here.

In the proof of Theorem 2.1, we take "K ample" as our definition of "$c_1(M)<0$".

Proof of Theorem 2.1. Let p be a positive integer such that K^p is very ample and let $\omega_0, \ldots, \omega_N$ be a basis for $H = H^0(M; K^p)$. Then the mapping

$$\iota\colon x \to (\omega_0(x), \ldots, \omega_N(x)) \qquad x \in M$$

defines an imbedding of M into $P_N(\mathbf{C})$. (Although $\omega_0(x), \ldots, \omega_N(x)$, are not numbers, their ratio makes sense and defines a point of $P_N(\mathbf{C})$). Every holomorphic transformation φ of M induces a linear transformation of H which will be denoted by $\rho(\varphi)$. We denote by $\sigma(\varphi)$ the projective transformation of $P_N(\mathbf{C})$ induced by $\rho(\varphi)$. Then

$$\sigma(\varphi) \circ \iota = \iota \circ \varphi .$$

In other words, the imbedding $\iota\colon M \to P_N(\mathbf{C})$ allows us to represent the group $\mathfrak{H}(M)$ of holomorphic transformations by a group of projective transformations of $P_N(\mathbf{C})$. It is clear that both ρ and σ are faithful representations.

Lemma 1. *The image $\sigma(\mathfrak{H}(M))$ of σ consists of exactly those projective transformations of $P_N(\mathbf{C})$ which preserve $\iota(M)$.*

Proof of Lemma 1. Let T be a projective transformation of $P_N(\mathbf{C})$ which preserves $\iota(M)$. Let φ be the restriction of T to $\iota(M)$. Since ι is an imbedding, φ can be considered as a holomorphic transformation of M. Now it suffices to show that $T = \sigma(\varphi)$. Since $\sigma(\varphi) \cdot T^{-1}$ is a projective transformation of $P_N(\mathbf{C})$ which induces the identity transformation on $\iota(M)$, we have only to show that *if T is a projective transformation of $P_N(\mathbf{C})$ which induces the identity transformation of $\iota(M)$, then T is the identity transformation of $P_N(\mathbf{C})$.* Let T be such a projective transformation and τ a linear transformation of H which induces T. We shall show that $\tau = cI$, where c is a constant and I is the identity transformation of H. From $T \circ \iota = \iota$ and from the definition of ι, it follows that

$$(\tau\omega)(z) = c(z) \cdot \omega(z) \qquad \text{for } \omega \in H \quad \text{and} \quad z \in M,$$

where $c(z)$ is a nonzero complex number which is independent of ω. Since both ω and $\tau\omega$ are holomorphic sections, $c(z)$ must be also holomorphic in z. As M is compact, $c(z)$ must be constant. This completes the proof of Lemma 1.

Lemma 1 implies that $\sigma(\mathfrak{H}(M))$ is a closed subgroup of the projective transformation group of $P_N(\mathbf{C})$. It shows also that $\sigma(\mathfrak{H}(M))$ is an algebraic group.

We shall now construct a bounded domain in H which is invariant by the group $\rho(\mathfrak{H}(M))$. To this end we introduce a real valued function v on H, which is very much like a norm. Every holomorphic section of the canonical line bundle K is a holomorphic n-form on M. Hence, every element ω of $H=H^0(M; K^p)$ can be symbolically written locally as follows:

$$\omega=f\cdot(dz^1\wedge\cdots\wedge dz^n)^p,$$

where f is a holomorphic function defined in the coordinate neighborhood in which $z^1, \ldots, z^n$ are valid. We define

$$v(\omega)=\int_M i^{n^2}\cdot(f\bar{f})^{1/p}\cdot dz^1\wedge\cdots\wedge dz^n\wedge d\bar{z}^1\wedge\cdots\wedge d\bar{z}^n.$$

Then $v(\omega)$ is well defined, independently of the choice of local coordinate system. The following lemma is trivial.

Lemma 2. (1) $v(\omega)\geqq 0$, *and* $v(\omega)=0$ *if and only if* $\omega=0$;

(2) $v(c\,\omega)=|c|^{2/p}\,v(\omega)$ *for* $c\in\mathbf{R}$;

(3) v *is a continuous function on the finite dimensional vector space* H;

(4) $v(\varphi^*\omega)=v(\omega)$ *for* $\varphi\in H(M)$.

We define now a bounded domain in H.

Lemma 3. *The open subset* D *of* H *defined by*

$$D=\{\omega\in H;\ v(\omega)<1\}$$

is a star-like bounded domain invariant by $\mathfrak{H}(M)$.

Proof of Lemma 3. By (2) of Lemma 2, every point of D can be joined to the origin by a straight line in D, showing that D is star like and, in particular, connected. To see that D is bounded, let $\omega_0, \ldots, \omega_N$ be any basis for H. Let S^{2N+1} be the unit sphere in H defined by

$$S^{2N+1}=\{\textstyle\sum a_i\,\omega_i;\ \sum|a_i|^2=1\}.$$

Let v_0 be the minimum value of the function v on S^{2N+1}; since v is continuous, v_0 exists and, by (1) of Lemma 2, must be positive. Let r be a positive number such that $r^{2/p}\,v_0>1$. Then, by (2) of Lemma 2, D is contained in the ball B defined by

$$B=\{\textstyle\sum a_i\,\omega_i;\ \sum|a_i|^2<r^2\}.$$

Finally, the invariance of D by $\rho(H(M))$ follows from (4) of Lemma 2. This completes the proof of Lemma 3.

We shall now complete the proof of Theorem 2.1. Let $\mathfrak{G}$ be the group of linear transformations of H leaving the bounded domain D invariant, where D is defined in Lemma 3. From Theorem 1.2 it follows that $\mathfrak{G}$ is a compact Lie group. On the other hand, since $\mathfrak{H}(M)$ is a complex Lie group and the action $\mathfrak{H}(M)\times M\to M$ is holomorphic (by Theorem 1, 1), the representation $\rho\colon \mathfrak{H}(M)\to \mathrm{GL}(N+1;\mathbf{C})$ is holomorphic. Since $\rho(\mathfrak{H}(M))$ is contained in the compact subset $\mathfrak{G}$ of $\mathrm{GL}(N+1;\mathbf{C}) \subset \mathbf{C}^{(N+1)^2}$, ρ maps the identity component of $\mathfrak{H}(M)$ into the identity element of $\mathrm{GL}(N+1;\mathbf{C})$. Since ρ is faithful, $\mathfrak{H}(M)$ is discrete. If we denote the natural homomorphism $\mathrm{GL}(N+1;\mathbf{C})\to \mathrm{PGL}(N;\mathbf{C})$ by π, then $\sigma=\pi\circ\rho$ and $\sigma(\mathfrak{H}(M))$ is a closed subgroup of the compact group $\pi(\mathfrak{G})$ by Lemmas 1 and 3. Hence, $\sigma(\mathfrak{H}(M))$ is compact. Since $\mathfrak{H}(M)$ is discrete and σ is faithful, $\mathfrak{H}(M)$ is finite. q.e.d.

Remark. In the course of the proof, we have established that *if M is a compact complex manifold, the identity component of $\mathfrak{H}(M)$ leaves every holomorphic section of K^p, $(p\geqq 0)$, fixed* (whether K is ample or not).

The second generalization is the following (Kobayashi [9, 10], Wu [1]).

Theorem 2.2. *Let M be a compact hyperbolic manifold. Then the group $\mathfrak{H}(M)$ of holomorphic transformations of M is finite.*

Proof. Let d_M be the intrinsic distance defined in § 1. As we saw in the proof of Theorem 1.4, the group $\mathfrak{H}(M)$ is a closed subgroup of the group $\mathfrak{J}(M)$ of isometries of M with respect to the distance d_M. Since M is compact, $\mathfrak{J}(M)$ is compact (see Theorem 1.1 of Chapter II) and hence $\mathfrak{H}(M)$ is also compact. It suffices therefore to prove that the identity component of $\mathfrak{H}(M)$ reduces to the identity element. Assume that $\dim \mathfrak{H}(M)$ is positive. Since $\mathfrak{H}(M)$ is a complex Lie group by Theorem 1.1, it is generated by complex 1-parameter subgroups. It suffices therefore to show that the group $\mathbf{C}$ cannot act holomorphically on M except in a trivial manner. Let $f\colon \mathbf{C}\times M\to M$ be a holomorphic action of $\mathbf{C}$ on M. For each fixed $p\in M$, the mapping $a\in\mathbf{C}\to f(a,p)\in M$ is holomorphic and hence is distance-decreasing with respect to $d_{\mathbf{C}}$ and d_M (see Lemma in the proof of Theorem 1.4). Since $d_{\mathbf{C}}$ is identically equal to zero, this means that $d_M(f(0,p), f(a,p))=0$ for all elements $a\in\mathbf{C}$. Since $f(0,p)=p$ and d_M is a distance, we may conclude that $f(a,p)=p$ for all $a\in\mathbf{C}$. q.e.d.

Examples. If M is a compact Kähler manifold with negative definite Ricci tensor, then $c_1(M)$ is negative and, by Theorem 2.1, the group $\mathfrak{H}(M)$ of holomorphic transformations is finite. On the other hand, if M is a compact hermitian manifold with negative holomorphic sectional curvature, then M is hyperbolic (Kobayashi [9, 10]) and, by Theorem 2.2, $\mathfrak{H}(M)$ is finite. If M is a complete intersection submanifold of r non-singular hypersurfaces of degrees $a_1, \dots, a_r$ in $P_{n+r}(\mathbf{C})$ such that $n+r+1$

$>a_1+\cdots+a_r$, then $c_1(M)$ is negative (see Hirzebruch [1; p. 159]) and, by Theorem 2.1, the group $\mathfrak{H}(M)$ is finite. In particular, if M is a non-singular hypersurface of degree greater than $n+2$ in $P_{n+1}(\mathbf{C})$, then $\mathfrak{H}(M)$ is finite. It is of some interest to note that the hypersurface M in $P_{n+1}(\mathbf{C})$ defined by $\sum_{i=0}^{n+1}(z^i)^d=0$ in terms of a homogeneous coordinate system $z^0, \ldots, z^{n+1}$ is not hyperbolic for any degree d, provided $n \geqq 2$. In fact, such a manifold contains a rational curve:

$$(u, v) \in P_1(\mathbf{C}) \to (u, v, \omega u, \omega v, 0, \ldots, 0) \in P_{n+1}(\mathbf{C}),$$

where ω denotes a d-th root of -1. On the other hand, I know of no example of a compact hyperbolic manifold whose first Chern class is not negative.

If M is of the form D/Γ, where D is a bounded domain in $\mathbf{C}^n$ and Γ is a properly discontinuous group of holomorphic transformations acting freely on D, then M is hyperbolic and also $c_1(M)$ is negative. For the proof of the first assertion, see Kobayashi [9, 10]. Let $2\sum g_{\alpha\bar{\beta}}\, dz^\alpha\, d\bar{z}^\beta$ be the Bergman metric of D. Since it is invariant by Γ, it may be considered also as a metric on $M=D/\Gamma$. Similarly, the 2-form $-\frac{1}{2\pi i}\sum g_{\alpha\bar{\beta}}\, dz^\alpha \wedge d\bar{z}^\beta$ may be considered as a 2-form on M. From the definition of the Bergman metric, it is clear that this 2-form represents the first Chern class $c_1(M)$. Since $(g_{\alpha\bar{\beta}})$ is positive definite, $c_1(M)$ is negative, thus proving the second assertion. The holomorphic transformation group $\mathfrak{H}(M)$ of $M=D/\Gamma$ is therefore finite either by Theorem 2.1 or by Theorem 2.2. The finiteness of $\mathfrak{H}(M)$ for $M=D/\Gamma$ has been proved by Bochner [2], Hawley [1], Sampson [1].

In connection with Theorem 2.1 and one of the examples above, we mention the following result.

Theorem 2.3. *Let M be a non-singular hypersurface of degree d in $P_{n+1}(\mathbf{C})$. If $n \geqq 2$ and $d \geqq 3$, then the group $\mathfrak{H}(M)$ of holomorphic transformations of M is finite, except in the case when $n=2$, $d=4$.*

See Matsumura and Monsky [1], where a completely algebraic proof is given. Lemma 14.2 in Kodaira-Spencer [1] shows also that $\dim \mathfrak{H}(M)=0$ if $n \geqq 2$ and $d \geqq 3$. Matsumura and Monsky show that $\mathfrak{H}(M)$ can be an infinite discrete group when $n=2$ and $d=4$.

The reader will find also a completely algebraic proof of Theorem 2.1 in Matsumura [1].

We say that an algebraic manifold M of dimension n is of *general type* if

$$\sup \lim_{m \to +\infty} \frac{1}{m^n} \dim H^0(M, K^m) > 0,$$

where K denotes the canonical line bundle of M. The following theorem generalizes Theorem 2.1.

Theorem 2.4. *If M is a projective algebraic manifold of general type, then its group $\mathfrak{H}(M)$ of holomorphic transformations is finite.*

For a completely algebraic proof of this theorem, see Matsumura [1]. A transcendental proof can be also given along the same line as the proof of Theorem 2.1. We again map $\mathfrak{H}(M)$ onto a group of linear transformations of the vector space $H=H^0(M, K^p)$ leaving a certain star-like bounded domain D invariant. The only nontrivial part of the proof is to show that this representation is faithful if p is large. But this follows from the result of Kodaira to the effect that we can obtain a projective imbedding of M using a certain subspace of $H=H^0(M, K^p)$ for p large. (For the detail, we refer the reader to the Addendum in Kobayashi-Ochiai [2].)

For a compact Riemann surface we have the following very precise result of Hurwitz [1].

Theorem 2.5. *Let M be a compact Riemann surface of genus $p \geqq 2$. Then the order of the group of holomorphic transformations of M is at most $84(p-1)$.*

We shall only indicate an outline of the proof. Let V be a compact Riemann surface of genus p' and $f\colon M \to V$ an n-fold covering projection with branch points. Let $a \in M$ be a branch point. With respect to a local coordinate system z with origin at a and a local coordinate system w with origin at $f(a)$, the mapping f is given locally by $w=z^m$ around a. Then $m-1$ is called the degree of ramification of f at a. Let $a_1, \dots, a_k$ be the branch points of f with degrees of ramification $m_1, \dots, m_k$. Then the *Riemann-Hurwitz relation states*

$$\chi(M)+\sum_{i=1}^{k} m_i = n \cdot \chi(V),$$

where $\chi(M)$ and $\chi(V)$ denote the Euler numbers of M and V. This formula can be easily verified by taking a triangulation of V such that $f(a_1), \dots, f(a_k)$ are vertices and the induced triangulation of M and then by counting the numbers of vertices, edges and faces.

Let $\mathfrak{G}$ be a finite group of holomorphic transformations of M; we know already that the group of holomorphic transformations of M is finite if the genus p of M is greater than 1. Let $\mathfrak{G}_a$ denote the isotropy subgroup of $\mathfrak{G}$ at $a \in M$. Let $f\colon M \to M/\mathfrak{G}$ be the natural projection. If the order m of $\mathfrak{G}_a$ is greater than 1 and if z is a local coordinate system around $a \in M$, then we introduce a local coordinate system w around

$f(a) \in M/\mathfrak{G}$ by $z = w^m$. In this way, $M/\mathfrak{G}$ becomes a compact Riemann surface which we shall denote by V. Then M is a branched covering of V with projection f, to which we apply the Riemann-Hurwitz relation. The degree of ramification of f at a is equal to $m-1$. If we denote by n the order of $\mathfrak{G}$, then the $\mathfrak{G}$-orbit through a consists of n/m points. The sum of the degrees of ramification of f at these points on the $\mathfrak{G}$-orbit of a is therefore equal to $\frac{n}{m}(m-1) = n\left(1-\frac{1}{m}\right)$. Hence, the Riemann-Hurwitz relation is of the form

$$\chi(M) + n \sum_{i=1}^{k} \left(1 - \frac{1}{m_i}\right) = n \cdot \chi(V).$$

Since m_i is the order of a subgroup of $\mathfrak{G}$, m_i divides n. If we denote by p' the genus of V, then the formula above may be rewritten as follows:

$$\frac{2p-2}{n} = 2p' - 2 + \sum_{i=1}^{k} \left(1 - \frac{1}{m_i}\right).$$

If $p' \geqq 2$, then $(p-1)/n \geqq 1$. If $p' = 1$, then $(2p-2)/n \geqq \left(1 - \frac{1}{m_1}\right) \geqq \frac{1}{2}$ and hence $n \leqq 4(p-1)$. Finally, consider the case $p' = 0$. Then

$$(2p-2)/n = -2 + \sum_{i=1}^{k} \left(1 - \frac{1}{m_i}\right) = k - 2 - \sum_{i=1}^{k} \frac{1}{m_i}.$$

It follows that $k \geqq 3$. If $k \geqq 5$, then $(2p-2)/n \geqq \frac{k}{2} - 2 \geqq \frac{1}{2}$ and $n \leqq 4(p-1)$. For $k = 4$, we have the following possibilities:

m_1	m_2	m_3	m_4	$(2p-2)/n$	
>2	>2	>2	>2	$\geqq \frac{2}{3}$	$n \leqq 3(p-1)$
$=2$	>2	>2	>2	$\geqq \frac{1}{2}$	$n \leqq 4(p-1)$
$=2$	$=2$	>2	>2	$\geqq \frac{1}{3}$	$n \leqq 6(p-1)$
$=2$	$=2$	$=2$	>2	$\geqq \frac{1}{6}$	$n \leqq 12(p-1)$

For $k = 3$, we have the following possibilities:

m_1	m_2	m_3	$(2p-2)/n$	
>3	>3	>3	$\geqq \frac{1}{4}$	$n \leqq 8(p-1)$
$=3$	>3	>3	$\geqq \frac{1}{6}$	$n \leqq 12(p-1)$
$=3$	$=3$	>3	$\geqq \frac{1}{12}$	$n \leqq 24(p-1)$
$=2$	>4	>4	$\geqq \frac{1}{10}$	$n \leqq 20(p-1)$
$=2$	$=4$	>4	$\geqq \frac{1}{20}$	$n \leqq 40(p-1)$
$=2$	$=3$	>3	$\geqq \frac{1}{42}$	$n \leqq 84(p-1)$

Let M be a compact complex manifold and K be its canonical line bundle. Let k be a positive integer such that K^k is very ample over some nonempty open set U of M in the following sense. At each point x of M, let $H(x)$ be the subspace of $H = H^0(M; K^k)$ consisting of holomorphic sections of K^k vanishing at x. Assume that, for each $x \in U$, $H(x)$ is a hyperplane of H and that the mapping $x \in U \to H(x)$ gives an imbedding of U into the projective space $P(H^*)$ (see the proof of Theorem 2.1). Then the natural representation ρ of $\mathfrak{H}(M)$ on H is faithful. In fact, if a holomorphic transformation of M leaves every holomorphic section of K^k fixed, then it leaves every point of U fixed and, being holomorphic, it leaves every point of M fixed. In particular, if M is a compact Riemann surface of genus $p \geqq 2$, then K is very ample over some nonempty open subset U, and it follows that a holomorphic transformation of M leaving every holomorphic 1-form fixed is the identity transformation. Thus we have the following result of Hurwitz [1]:

Theorem 2.6. *A holomorphic transformation of a compact Riemann surface of genus $p \geqq 2$ is the identity transformation if it induces the identity transformation on the first homology group $H_1(M; \mathbf{R})$.*

For higher dimensional analogs of Theorem 2.6, see Theorem 4.4 of Chapter II and Borel-Narasimhan [1].

For more results on automorphisms of compact Riemann surfaces of genus $\geqq 2$, see Macbeath [1], Lehner-Newman [1], Accola [1, 2], Lewittes [1].

For an analog of Theorem 2.5 for algebraic surfaces, see Andreotti [1].

Somewhat related with the results of this section is the following theorem of Gottschling [1]. Let H_m be the Siegel upper-half space of degree m, i.e., the space of complex symmetric matrices of degree m with positive definite imaginary part and Γ_m be the Siegel modular group of degree m. Then, for $m \geqq 3$, the group of holomorphic transformations of H_m/Γ_m consists of the identity element only.

3. Holomorphic Vector Fields and Holomorphic 1-Forms

If Z is a holomorphic vector field and ω is a holomorphic 1-form on a complex manifold M, then $\omega(Z)$ is a holomorphic function on M. If M is compact, this function must be constant. This simple fact yields some useful results.

Proposition 3.1. *Let M be a compact complex manifold and $\mathfrak{h} = \mathfrak{h}(M)$ the Lie algebra of holomorphic vector fields on M. Denote by $\mathscr{C}^{1,0}$ the space of closed holomorphic 1-forms on M. Define*

$$\mathfrak{h}_1 = \{Z \in \mathfrak{h};\ \omega(Z) = 0 \text{ for all } \omega \in \mathscr{C}^{1,0}\}.$$

Then

(1) $\mathfrak{h}_1$ *is an ideal of* $\mathfrak{h}$ *and contains* $[\mathfrak{h}, \mathfrak{h}]$;

(2) $\dim \mathfrak{h}/\mathfrak{h}_1 \leqq \frac{1}{2} b_1$,

where b_1 *is the first Betti number of* M.

Proof. We recall the following general formula relating a 1-form ω and vector fields Z and W:

$$2d\omega(Z, W) = Z(\omega(W)) - W(\omega(Z)) - \omega([Z, W]).$$

(1) follows immediately from this formula.

Let $\mathscr{C}^{0,1} = \bar{\mathscr{C}}^{1,0}$ be the space of closed anti-holomorphic 1-forms. It is a simple matter to verify that $\mathscr{C}^{1,0}$ (resp. $\mathscr{C}^{0,1}$) is the space of closed (1, 0)-forms (resp. (0, 1)-forms). Let $\mathscr{C}^1$ be the space of closed (complex) 1-forms. Then

$$\mathscr{C}^{1,0} + \mathscr{C}^{0,1} \subset \mathscr{C}^1.$$

Let $\mathscr{B}^1$ be the space of cobounding 1-forms, i.e., 1-forms of the type df. Then

$$(\mathscr{C}^{1,0} + \mathscr{C}^{0,1})/\mathscr{B}^1 \cap (\mathscr{C}^{1,0} + \mathscr{C}^{0,1}) \subset \mathscr{C}^1/\mathscr{B}^1 = H^1(M; \mathbf{C}).$$

Let

$$\mathscr{D}^{1,0} = \{\omega \in \mathscr{C}^{1,0};\ \omega(Z) = 0 \text{ for all } Z \in \mathfrak{h}\},$$

$$\mathscr{D}^{0,1} = \{\bar{\omega} \in \mathscr{C}^{0,1};\ \bar{\omega}(\bar{Z}) = 0 \text{ for all } Z \in \mathfrak{h}\} = \bar{\mathscr{D}}^{1,0}.$$

Then the pairing

$$(\omega, Z) \in \mathscr{C}^{1,0} \times \mathfrak{h} \to \omega(Z) \in \mathbf{C}$$

induces a dual pairing between $\mathscr{C}^{1,0}/\mathscr{D}^{1,0}$ and $\mathfrak{h}/\mathfrak{h}_1$.

We shall show

$$\mathscr{B}^1 \cap (\mathscr{C}^{1,0} + \mathscr{C}^{0,1}) \subset \mathscr{D}^{1,0} + \mathscr{D}^{0,1}.$$

Let $df = \alpha + \bar{\beta} \in \mathscr{B}^1 \cap (\mathscr{C}^{1,0} + \mathscr{C}^{0,1})$, where $\alpha, \beta \in \mathscr{C}^{1,0}$. Then

$$df + d\bar{f} = \alpha + \beta + \bar{\alpha} + \bar{\beta}.$$

Let $Z \in \mathfrak{h}$. Then

$$(df + d\bar{f})(Z) = (\alpha + \beta)(Z).$$

Since α and β are holomorphic, the right hand side is constant. On the other hand, the left hand side vanishes at the maximum point of the real valued function $f + \bar{f}$. Hence,

$$(\alpha + \beta)(Z) = 0.$$

Similarly, from

$$(df - d\bar{f})(Z) = (\alpha - \beta)(Z),$$

we obtain

$$(\alpha - \beta)(Z) = 0.$$

Hence, $\alpha(Z) = \beta(Z) = 0$. This proves our assertion that df is in $\mathscr{D}^{1,0} + \mathscr{D}^{0,1}$.

We may now conclude

$$\begin{aligned} 2\cdot \dim \mathfrak{h}/\mathfrak{h}_1 &= 2\cdot \dim \mathscr{C}^{1,0}/\mathscr{D}^{1,0} = \dim(\mathscr{C}^{1,0}+\mathscr{C}^{0,1})/(\mathscr{D}^{1,0}+\mathscr{D}^{0,1}) \\ &\leqq \dim(\mathscr{C}^{1,0}+\mathscr{C}^{0,1})/\mathscr{B}^1 \cap (\mathscr{C}^{1,0}+\mathscr{C}^{0,1}) \\ &\leqq \dim \mathscr{C}^1/\mathscr{B}^1 = b_1. \qquad \text{q.e.d.} \end{aligned}$$

Remark. A holomorphic vector field Z with non-empty zero set belongs to the ideal $\mathfrak{h}_1$.

The ideal $\mathfrak{h}_1$ of $\mathfrak{h}(M)$ was introduced by Lichnerowicz [3] to study $\mathfrak{h}(M)$ of a compact Kähler manifold M with non-positive, non-negative or zero first Chern class. See also Matsushima [5].

4. Holomorphic Vector Fields on Kähler Manifolds

Let M be a Kähler manifold of dimension n. Let Z be a complex vector field with components $\zeta^\alpha, \zeta^{\bar\alpha}$ in terms of a local coordinate system $z^1, \dots, z^n$, i.e.,

$$Z = \sum \zeta^\alpha \frac{\partial}{\partial z^\alpha} + \sum \zeta^{\bar\alpha} \frac{\partial}{\partial \bar z^\alpha}.$$

We can write

$$\nabla Z = \nabla' Z + \nabla'' Z,$$

where $\nabla' Z$ and $\nabla'' Z$ are defined by the property that

$$\nabla'_{\bar W} Z = 0 \quad \text{and} \quad \nabla''_W Z = 0 \qquad \text{for all vectors } W \text{ of type } (1,0).$$

In terms of a local coordinate system,

$$\nabla' Z = \sum \nabla_\beta \zeta^\alpha dz^\beta \otimes \frac{\partial}{\partial z^\alpha} + \sum \nabla_\beta \zeta^{\bar\alpha} dz^\beta \otimes \frac{\partial}{\partial \bar z^\alpha},$$

$$\nabla'' Z = \sum \nabla_{\bar\beta} \zeta^\alpha d\bar z^\beta \otimes \frac{\partial}{\partial z^\alpha} + \sum \nabla_{\bar\beta} \zeta^\alpha d\bar z^\beta \otimes \frac{\partial}{\partial \bar z^\alpha}.$$

Similarly, for any tensor field K, we can write

$$\nabla K = \nabla' K + \nabla'' K.$$

Given a complex vector field Z of type (1, 0), we denote the (0, 1)-form corresponding to Z by ζ. In terms of their components, we have

$$Z = \sum \zeta^\alpha \frac{\partial}{\partial z^\alpha} \leftrightarrow \zeta = \sum \zeta_{\bar\beta} d\bar z^\beta \quad \text{with} \quad \zeta_{\bar\beta} = \sum g_{\alpha\bar\beta} \zeta^\alpha.$$

Proposition 4.1. *A complex vector field Z of type* (1, 0) *on a Kähler manifold M is holomorphic if and only if $\nabla'' Z = 0$, or equivalently, $\nabla'' \zeta = 0$ (where ζ is the* (0, 1)*-form corresponding to Z).*

Proof. Clearly, Z is holomorphic if and only if, for each point p of M, there exists a local coordinate system $z^1, \ldots, z^n$ around p such that $(\partial\zeta^\alpha/\partial\bar{z}^\beta)_p = 0$. On the other hand, since M is a Kähler manifold, for each point p there exists a local coordinate system $z^1, \ldots, z^n$ around p such that $(\nabla_{\bar\beta})_p = (\partial/\partial\bar{z}^\beta)_p$ (i.e., such that the Christoffel symbols vanish at p). Hence, Z is holomorphic if and only if $\nabla_{\bar\beta}\zeta^\alpha = 0$. q.e.d.

Theorem 4.2. *Let M be a Kähler manifold, Z a complex vector field of type* (1, 0) *and ζ the corresponding* (0, 1)*-form. If Z is holomorphic, then*

$$\tfrac{1}{2}\Delta\zeta = \Delta''\zeta = \sum R_{\alpha\bar\beta}\zeta^\alpha\, d\bar{z}^\beta.$$

Conversely, if M is compact and

$$\int_M g(\Delta''\zeta - \sum R_{\alpha\bar\beta}\zeta^\alpha\, d\bar{z}^\beta, \zeta)\, dv = 0,$$

then Z is holomorphic.

Proof. In Appendix 3, the following general formula for a (0, 1)-form is proved:

$$\Delta''\zeta = \sum(-\nabla_\alpha\nabla^\alpha\zeta_{\bar\beta}\, d\bar{z}^\beta + R_{\alpha\bar\beta}\ \zeta^\alpha\, d\bar{z}^\beta).$$

Since

$$\nabla^\alpha\zeta_{\bar\beta} = \sum g^{\alpha\bar\gamma} g_{\delta\bar\beta}\nabla_{\bar\gamma}\zeta^\delta,$$

Proposition 4.1 implies $\nabla^\alpha\zeta_{\bar\beta} = 0$ if X is holomorphic. Hence, we obtain the first statement of the theorem. To prove the converse, we use the following integral formula (see Theorem 4 of Appendix 2) expressed in terms of a local coordinate system:

$$\int_M \{-\sum(\Delta''\zeta)_{\bar\beta}\bar\zeta^\beta + \sum R_{\alpha\bar\beta}\zeta^\alpha\bar\zeta^\beta + \sum\nabla_{\bar\beta}\zeta^\alpha\cdot\overline{\nabla^\beta\zeta_{\bar\alpha}}\}\, dv = 0.$$

The first two terms of the integrand cancel each other by our assumption. Hence,

$$\int_M \sum\nabla_{\bar\beta}\zeta^\alpha\cdot\overline{\nabla^\beta\zeta_{\bar\alpha}}\, dv = 0.$$

This implies $\nabla_{\bar\beta}\zeta^\alpha = 0$. By Proposition 4.1, Z is holomorphic. q.e.d.

The first half of the theorem is due to Bochner [1] and its converse to Yano [4].

Theorem 4.3. *Let M be a compact Kähler manifold and $Z = X - iJX$ a complex vector field of type* (1, 0) *with real part X. Then X is an infinitesimal isometry if and only if Z is holomorphic and* $\operatorname{div} X = 0$.

Proof. We use the characterization for an infinitesimal isometry obtained in Theorem 2.3 of Chapter II. By Theorem 4.2, Z is holomorphic if and only if X satisfies (1) in Theorem 2.3 of Chapter II. q.e.d.

Theorem 4.3 is due to Yano [4].

Theorem 4.4. *Let M be a compact Kähler manifold and Z a holomorphic vector field (of type $(1,0)$) with the corresponding $(0,1)$-form ζ. Then*

(1) *$\zeta = H\zeta + d''f$, where $H\zeta$ is the harmonic part of ζ and f is a (complex valued) function. Such a function f is unique if it is normalized by the property $\int_M f\,dv = 0$.*

(2) *$\zeta = d''f$ if and only if $\alpha(Z)=0$ for every holomorphic 1-form α, i.e., if and only if $Z \in \mathfrak{h}_1$, where $\mathfrak{h}_1$ is the ideal introduced in § 3.*

(3) *The real part X of Z is an infinitesimal isometry if and only if the real part of f is a constant. (This means that if f is normalized as in (1), then f is purely imaginary.)*

Proof. By Proposition 4.1, $d''\zeta = 0$. Now, (1) follows from the Hodge-Kodaira decomposition theorem:

$$\zeta = H\zeta + d''\delta''\varphi + \delta'' d''\varphi,$$

where φ is a form with the same bidegree as ζ. If $\zeta = H\zeta + d''f = H\zeta + d''g$, then $d''(f-g)=0$, that is, $f-g$ is holomorphic and hence a constant.

To prove (2), we observe first that if α is a holomorphic 1-form, then $\alpha(Z)$ is a holomorphic function on M and hence is a constant. It suffices therefore to prove that the integral $\int_M \alpha(Z)\,dv$ vanishes for all holomorphic 1-forms α if and only if $H\zeta=0$. Assume $H\zeta=0$. Then $\alpha(Z)=g(\alpha,\zeta)= g(\alpha, d''f)$, where g denotes the inner product on the cotangent spaces defined by the metric. Since α is holomorphic, it is harmonic. Since a harmonic form is perpendicular to $d'\bar{f} = \overline{d''f}$ in the Hodge-Kodaira decomposition, the integral $\int_M g(\alpha, d''f)\,dv$ vanishes. Assume, conversely, that $\alpha(Z)=0$ and let $\alpha = \overline{H\zeta}$; since $H\zeta$ is a harmonic $(0,1)$-form, its complex conjugate is a harmonic $(1,0)$-form and hence is holomorphic. Then

$$0 = \int_M \overline{H\zeta}(Z)\,dv = \int_M (g(\overline{H\zeta}, H\zeta) + g(\overline{H\zeta}, d''f))\,dv = \int_M g(\overline{H\zeta}, H\zeta)\,dv.$$

Hence, $H\zeta = 0$.

X is an infinitesimal isometry if and only if $\operatorname{div} X = 0$ (Theorem 4.3). On the other hand, $\operatorname{div} X = 0$ if and only if $\delta(\zeta+\bar{\zeta})=0$. But

$$\begin{aligned}\delta(\zeta+\bar{\zeta}) &= \delta''\zeta + \delta'\bar{\zeta} = \delta'' d'' f + \delta' d'\bar{f} \\ &= (\delta'' d'' + d''\delta'')f + (\delta' d' + d'\delta')\bar{f} \\ &= \Delta'' f + \Delta'\bar{f} = \tfrac{1}{2}\Delta f + \tfrac{1}{2}\Delta\bar{f} = \tfrac{1}{2}\Delta(f+\bar{f}).\end{aligned}$$

Since $\Delta(f+\bar{f})=0$ if and only if $f+\bar{f}$ is a constant, we obtain the assertion in (3). q.e.d.

(1) and (2) of Theorem 4.4 are due to Lichnerowicz [6].

Corollary 4.5. *In Theorem 4.4, if the zero set of Z is nonempty, then $\zeta=d''f$.*

Proof. If α is a holomorphic 1-form, then $\alpha(Z)$ is constant. If the zero set of Z is nonempty, then $\alpha(Z)=0$. Our assertion follows from (2) of Theorem 4.4. q.e.d.

Corollary 4.6. *In Theorem 4.4, assume that the real part X of Z is an infinitesimal isometry. Let ζ be the $(0,1)$-form corresponding to Z and ξ be the real 1-form corresponding to X. Then the following statements are mutually equivalent:*

(1) *The zero set of Z ($=$the zero set of X) is nonempty;*

(2) *$\zeta=d''f$, where f is a function with purely imaginary values;*

(3) *$\xi=Jdu$, where u is a real valued function.*

5. Compact Einstein-Kähler Manifolds

In this section we shall prove the following result of Matsushima [1].

Theorem 5.1. *Let M be a compact Einstein-Kähler manifold with non-zero Ricci tensor. Then the Lie algebra $\mathfrak{i}(M)$ of infinitesimal isometries is a real form of the Lie algebra $\mathfrak{h}(M)$ of holomorphic vector fields, i.e.,*

$$\mathfrak{h}(M)=\mathfrak{i}(M)+\sqrt{-1}\cdot\mathfrak{i}(M).$$

In the statement above, $\mathfrak{i}(M)$ is imbedded in $\mathfrak{h}(M)$ by identifying an infinitesimal isometry X with the corresponding holomorphic vector field $Z=X-iJX$ (see Theorem 4.3).

Proof. By our assumption, the Ricci tensor $R_{\alpha\bar{\beta}}$ and the metric tensor $g_{\alpha\bar{\beta}}$ satisfy the relation:

$$R_{\alpha\bar{\beta}}=c\cdot g_{\alpha\bar{\beta}}, \qquad \text{where } c \text{ is a nonzero constant.}$$

Let Z be a holomorphic vector field (of type $(1,0)$) and ζ the corresponding $(0,1)$-form. By Theorem 4.2,

$$\Delta''\zeta=c\zeta.$$

Substituting $\zeta=H\zeta+d''f$ (see Theorem 4.4, (1)) into this, we obtain

$$\Delta''d''f=c(H\zeta+d''f),$$

which shows that $H\zeta=0$. We may assume that f is normalized as in Theorem 4.4 and we write

$$f=u+iv,$$

where u and v are real valued functions. We shall show that

$$\Delta'' d'' u=c\, d'' u \quad \text{and} \quad \Delta'' d'' v=c\, d'' v.$$

Since $\zeta=d''f$ and $\Delta''\zeta=c\zeta$, we have

$$0=\Delta'' d'' f-c\, d'' f=d''\Delta'' f-d'' cf=d''(\Delta'' f-cf),$$

which shows that $\Delta'' f-cf$ is a holomorphic function and hence is a constant. But

$$\Delta'' f-cf=(\Delta'' u-cu)+i(\Delta'' v-cv)=(\tfrac{1}{2}\Delta u-cu)+i(\tfrac{1}{2}\Delta v-cv).$$

Hence, both the real part $(\Delta'' u-cu)$ and the imaginary part $(\Delta'' v-cv)$ of $\Delta'' f-cf$ must be also constant. It follows that $d''(\Delta'' u-cu)=0$ and $d''(\Delta'' v-cv)=0$, showing our assertion.

By Theorem 4.2, this means that the vector fields U and V of type (1, 0) corresponding to the (0, 1)-forms du'' and $i\,d''v$ are holomorphic. By (3) of Theorem 4.4, $-iU$ and V correspond to infinitesimal isometries since $-iu$ and iv are purely imaginary. Since $\zeta=d''f=d''u+i\,d''v$, we have

$$Z=i(-iU)+V.$$

This shows that $\mathfrak{h}(M)=\mathfrak{i}(M)+\sqrt{-1}\cdot\mathfrak{i}(M)$. q.e.d.

In the course of the proof, we established that $\Delta'' f-cf$ is a constant. Integrating $\Delta'' f-cf$ over M and observing that the integral of $\Delta'' f$ $(=\frac{1}{2}\Delta f)$ vanishes, we see that this constant is zero if f is normalized. Hence, $\Delta'' f=cf$, or

$$\Delta f=2cf.$$

Denote by $\mathscr{F}_{2c}$ the set of all complex valued functions f which are eigen functions of the Laplacian Δ with eigen value $2c$, i.e.,

$$\mathscr{F}_{2c}=\{f;\ \Delta f=2cf\}.$$

Then the correspondence

$$f\to\zeta=d''f\to Z$$

gives a complex linear isomorphism between $\mathscr{F}_{2c}$ and $\mathfrak{h}(M)$. The subspace of $\mathscr{F}_{2c}$ consisting of purely imaginary functions corresponds to $\mathfrak{i}(M)$.

Since $\mathfrak{h}(M)=0$ if $c<0$ by Theorem 2.1, Theorem 5.1 is of interest only when $c>0$.

Theorem 5.2. *Let M be a compact Kähler manifold with vanishing Ricci tensor. Then the Lie algebra $\mathfrak{h}(M)$ of holomorphic vector fields coincides with the Lie algebra $\mathfrak{i}(M)$ of infinitesimal isometries. It consists of parallel vector fields and is abelian.*

Proof. Let Z be a holomorphic vector field and ζ the corresponding $(0, 1)$-form. By Theorem 4.2,

$$\Delta'' \zeta = 0,$$

that is, ζ is harmonic. In the decomposition $\zeta = H\zeta + d''f$ in (1) of Theorem 4.4, the function f is zero (if it is normalized). By (3) of Theorem 4.4, the real part of Z is an infinitesimal isometry. This establishes $\mathfrak{h}(M) = \mathfrak{i}(M)$. By Corollary 4.2 of Chapter II, every infinitesimal isometry of M is a parallel vector field. Clearly, every parallel vector field is an infinitesimal isometry. Finally, the general formula

$$[X, Y] - (\nabla_X Y - \nabla_Y X) = 0$$

implies that the Lie algebra of parallel vector fields is abelian. q.e.d.

For a generalization of Matsushima's theorem to compact almost Einstein-Kähler manifolds, see Sawaki [1].

It is interesting to find out how large the class of compact Einstein-Kähler manifolds is. It is not known if there exists a non-homogeneous compact Einstein manifold. In this connection, see Berger [2] for a survey on Einstein manifolds and Aubin [1] for a construction of certain Einstein-Kähler metrics.

6. Compact Kähler Manifolds with Constant Scalar Curvature

In this section we shall prove a theorem of Lichnerowicz [2, 3] which generalizes Theorem 5.1 of Matsushima.

Theorem 6.1. *Let M be a compact Kähler manifold with constant scalar curvature. Let $\mathfrak{h}(M)$ denote the Lie algebra of holomorphic vector fields, $\mathfrak{i}(M)$ the Lie algebra of infinitesimal isometries (considered as a subalgebra of $\mathfrak{h}(M)$), $\mathfrak{c}$ the subalgebra of $\mathfrak{h}(M)$ consisting of parallel holomorphic vector fields and $\mathfrak{b}$ the ideal of $\mathfrak{h}(M)$ consisting of those vector fields Z such that $\alpha(Z) = 0$ for all holomorphic 1-forms α. Then*

(1) $\mathfrak{h}(M) = \mathfrak{b} + \mathfrak{c}$ *(Lie algebra direct sum);*

(2) $\mathfrak{b} = (\mathfrak{i}(M) \cap \mathfrak{b}) + \sqrt{-1}(\mathfrak{i}(M) \cap \mathfrak{b})$, *i.e.,* $\mathfrak{i}(M) \cap \mathfrak{b}$ *is a real form of* $\mathfrak{b}$*;*

(3) $\mathfrak{i}(M) = (\mathfrak{i}(M) \cap \mathfrak{b}) + \mathfrak{c}$.

Proof. We denote by Q the linear endomorphism on the space of $(0, 1)$-forms defined by

$$Q\zeta = \sum R_{\alpha\bar{\beta}} \zeta^{\alpha} d\bar{z}^{\beta} \qquad \text{for } \zeta = \sum \zeta_{\bar{\alpha}} d\bar{z}^{\alpha}.$$

Let Z be an arbitrary holomorphic vector field on M and ζ the corresponding (0, 1)-form. Following Theorem 4.4, we write

$$\zeta=\varphi+d''f, \qquad \text{where } \varphi=H\zeta=\text{the harmonic part of } \zeta.$$

As in § 5 we write

$$f=u+iv, \qquad \text{where } u \text{ and } v \text{ are real valued functions}.$$

We set

$$\xi=d''u \quad \text{and} \quad \eta=i d''v$$

so that

$$\zeta=\varphi+\xi+\eta.$$

We shall show that φ corresponds to a vector field belonging to $\mathfrak{c}$ and that ξ and η correspond to vector fields belonging to $\sqrt{-1}(\mathfrak{i}(M)\cap\mathfrak{b})$ and $\mathfrak{i}(M)\cap\mathfrak{b}$, respectively.

We shall make use of the following formula which follows easily from the second Bianchi identity:

Lemma. *If we denote by R the scalar curvature, then*

$$\sum\nabla^{\alpha}R_{\alpha\bar\beta}=\nabla_{\bar\beta}R \quad \text{and} \quad \sum\nabla^{\bar\beta}R_{\alpha\bar\beta}=\nabla_{\alpha}R.$$

Since φ is a harmonic form of degree (0, 1), its conjugate is holomorphic and hence $\nabla_{\beta}\varphi_{\bar\alpha}=0$. This fact, together with Lemma, implies

$$\Delta''(Q\varphi)=-\sum\nabla^{\bar\beta}(R_{\alpha\bar\beta}\varphi^{\alpha})=-\sum\varphi^{\alpha}\nabla_{\alpha}R=0.$$

Since Z is holomorphic, Theorem 4.2 implies

$$\Delta''\zeta-Q\zeta=0.$$

Since $\Delta''\varphi=0$, this implies

$$\Delta''(d''f)-Q(d''f)=Q\varphi.$$

Hence,

$$\int_M(\Delta''(d''f)-Q(d''f),\overline{d''f})\,dv=\int_M(Q\varphi,\overline{d''f})\,dv$$
$$=\int_M(\delta''(Q\varphi),\bar f)\,dv=0.$$

By Theorem 4.2, this means that the vector field corresponding to the (0, 1)-form $d''f$ is holomorphic. Hence, the vector field corresponding to φ $(=\zeta-d''f)$ is also holomorphic. By Proposition 4.1, $\nabla_{\bar\beta}\varphi_{\bar\alpha}=0$. Since we already know that $\nabla_{\beta}\varphi_{\bar\alpha}=0$, we can conclude that φ is parallel.

Since $d''f=\xi+\eta$ corresponds to a holomorphic vector field, Theorem 4.2 implies $\Delta''\xi-Q\xi=-(\Delta''\eta-Q\eta)$. Hence

$$\delta''(\Delta''\xi-Q\xi)=-\delta''(\Delta''\eta-Q\eta).$$

We shall show that both sides of this equality vanish by demonstrating that the left hand side is real and the right hand side is purely imaginary. We have

$$\begin{aligned}\delta''(\Delta''\xi - Q\xi) &= \delta'' d'' \delta'' \xi - \delta''(Q\xi)\\ &= \delta'' d'' \delta'' d'' u + \sum \nabla^{\bar\beta}(R_{\alpha\bar\beta}\cdot\nabla^{\alpha} u)\\ &= \Delta''\cdot\Delta'' u + \sum R_{\alpha\bar\beta}\cdot\nabla^{\bar\beta}\nabla^{\alpha} u.\end{aligned}$$

This shows that $\delta''(\Delta''\xi - Q\xi)$ is real. Similarly,

$$\delta''(\Delta''\eta - Q\eta) = i(\Delta''\Delta'' v + \sum R_{\alpha\bar\beta}\cdot\nabla^{\bar\beta}\nabla^{\alpha} v),$$

which shows that $\delta''(\Delta''\eta - Q\eta)$ is purely imaginary. Hence,

$$\delta''(\Delta''\xi - Q\xi) = \delta''(\Delta''\eta - Q\eta) = 0.$$

Now we have

$$\begin{aligned}\int_M (\Delta''\xi - Q\xi, \bar\xi)\, dv &= \int_M (\Delta''\xi - Q\xi, \overline{d'' u})\, dv\\ &= \int_M (\delta''(\Delta''\xi - Q\xi), \bar u)\, dv = 0.\end{aligned}$$

By Theorem 4.2, ξ corresponds to a holomorphic vector field. Hence, η corresponds also to a holomorphic vector field.

Denote by Z_0, Z_1, Z_2 the holomorphic vector fields corresponding to the (0, 1)-forms φ, ξ, η, respectively. Then

$$Z = Z_0 + Z_1 + Z_2.$$

We have shown already that Z_0 is in $\mathfrak{c}$. By (2) and (3) of Theorem 4.4, Z_1 is in $\sqrt{-1}(\mathfrak{i}(M)\cap\mathfrak{b})$ and Z_2 is in $\mathfrak{i}(M)\cap\mathfrak{b}$. The facts that

$$\mathfrak{b}\cap\mathfrak{c} = 0 \quad\text{and}\quad (\mathfrak{i}(M)\cap\mathfrak{b})\cap(\sqrt{-1}(\mathfrak{i}(M)\cap\mathfrak{b})) = 0$$

are also immediate consequences of Theorem 4.4.

To prove that $[\mathfrak{b}, \mathfrak{c}] = 0$, let

$$Z = \sum \nabla^{\alpha} f \frac{\partial}{\partial z^{\alpha}} \in \mathfrak{b} \quad\text{and}\quad W = \sum \varphi^{\alpha}\frac{\partial}{\partial z^{\alpha}} \in \mathfrak{c}.$$

Then

$$\begin{aligned}[Z, W] &= \sum (\nabla^{\alpha} f\cdot\nabla_{\alpha}\varphi^{\beta} - \varphi^{\beta}\nabla_{\beta}\nabla^{\alpha} f)\frac{\partial}{\partial z^{\alpha}} = -\sum (\varphi^{\beta}\nabla_{\beta}\nabla^{\alpha} f)\frac{\partial}{\partial z^{\alpha}}\\ &= -\sum (\varphi^{\beta}\nabla^{\alpha}\nabla_{\beta} f)\frac{\partial}{\partial z^{\alpha}} = -\sum \nabla^{\alpha}(\varphi^{\beta}\nabla_{\beta} f)\frac{\partial}{\partial z^{\alpha}}\end{aligned}$$

since W is parallel. But

$$\sum \overline{\varphi^\beta \nabla_\beta f} = \sum \overline{\varphi}_\beta \nabla^\beta \bar{f}.$$

Since $\sum \varphi_\beta\, d\bar{z}^\beta$ is a harmonic (0, 1)-form, $\sum \overline{\varphi}_\beta\, dz^\beta$ is a harmonic (1, 0)-form and hence is a holomorphic 1-form. We set $\alpha = \sum \overline{\varphi}_\beta\, dz^\beta$. On the other hand, the vector field $Z' = \sum \nabla^\beta \bar{f} \dfrac{\partial}{\partial z^\beta}$ is also in $\mathfrak{b}$. Since $\alpha(Z') = 0$ by Theorem 4.4, we obtain

$$\sum \overline{\varphi}_\beta \nabla^\beta \bar{f} = 0.$$

This completes the proof of the fact that $[Z, W] = 0$.

Finally, (3) follows from (1) and the fact that every element of $\mathfrak{c}$, being parallel, is in $\mathfrak{i}(M)$. q.e.d.

7. Conformal Changes of the Laplacian

In order to study compact Kähler manifolds with non-negative or non-positive first Chern class, it is convenient to introduce the Laplacian with respect to a hermitian metric conformal to the given Kähler metric. The results in this section are due to Lichnerowicz [6]. We follow both Lichnerowicz [6] and Matsushima [5].

Let M be a Kähler manifold and let e^σ be a real, positive function on M. We introduce operators δ''_σ and Δ''_σ operating on complex differential forms by

$$\delta''_\sigma \varphi = e^{-\sigma} \delta''(e^\sigma \cdot \varphi) \qquad \text{for every differential form } \varphi,$$

and

$$\Delta''_\sigma = \delta''_\sigma \circ d'' + d'' \circ \delta''_\sigma .$$

By a direct calculation using local coordinates (see Appendix 3), we obtain

$$\delta''_\sigma \varphi = \delta'' \varphi - \iota(d'\sigma)\, \varphi,$$

where $\iota(d'\sigma)$ denotes the interior product of φ with the complex vector field of type (0, 1) corresponding to the (1, 0)-form $d'\sigma$. We obtain easily

$$\Delta''_\sigma = \Delta'' - \left(d'' \circ \iota(d'\sigma) + \iota(d'\sigma) \circ d''\right).$$

If M is compact, we define a new inner product $(\ ,\ )_\sigma$ by

$$(\varphi, \bar{\psi})_\sigma = \int_M (\varphi, \bar{\psi}) \cdot e^\sigma \cdot dv.$$

Then

$$(d'' \varphi, \bar{\psi})_\sigma = (\varphi, \overline{\delta''_\sigma \psi})_\sigma .$$

In fact,

$$(d''\varphi, \bar{\psi})_\sigma = (d''\varphi, e^\sigma \bar{\psi}) = (\varphi, \overline{\delta''(e^\sigma \psi)}) = (\varphi, \overline{e^{-\sigma}\delta''(e^\sigma \psi)})_\sigma = (\varphi, \overline{\delta''_\sigma \psi})_\sigma.$$

A differential form φ is said to be *Δ''_σ-harmonic* if $\Delta''_\sigma \varphi = 0$.

Theorem 7.1. *Let M be a compact Kähler manifold and denote by $H^{p,q}_\sigma$ the space of Δ''_σ-harmonic (p, q)-forms and by Ω^p the sheaf of germs of holomorphic p-forms over M. Then*

$$\dim H^{p,q}_\sigma < \infty; \tag{1}$$

$$H^{p,q}_\sigma = H^q(M; \Omega^p). \tag{2}$$

In particular, $\dim H^{p,q}_\sigma$ *is independent of* σ.

The result (1) is due to Kodaira and (2) is the result of Dolbeault. For the proof and further references, see Hirzebruch [1; Chapter IV].

We define a tensor field with components $C_{\alpha\bar{\beta}}$ by

$$C_{\alpha\bar{\beta}} = R_{\alpha\bar{\beta}} - \nabla_{\bar{\beta}} \nabla_\alpha \sigma,$$

and denote by Q_σ the linear endomorphism on the space of $(0, 1)$-forms defined by

$$Q_\sigma \zeta = \sum C_{\alpha\bar{\beta}} \zeta^\alpha d\bar{z}^\beta \quad \text{for } \zeta = \sum \zeta_{\bar{\alpha}} d\bar{z}^\alpha.$$

Then the following theorem generalizes Theorem 4.2.

Theorem 7.2. *Let M be a Kähler manifold, Z a complex vector field of type $(1, 0)$ and ζ the corresponding $(0, 1)$-form. If Z is holomorphic, then*

$$\Delta''_\sigma \zeta - Q_\sigma(\zeta) = 0.$$

Conversely, if M is compact and

$$(\Delta''_\sigma \zeta - Q_\sigma(\zeta), \bar{\zeta})_\sigma = 0,$$

then Z is holomorphic.

Proof. We prove the following lemma which is a local formula.

Lemma. $\Delta''_\sigma \zeta - Q_\sigma(\zeta) = \Delta'' \zeta - Q(\zeta) - \nabla_S \zeta$, *where $Q = Q_0$ and S is the vector field of type $(0, 1)$ corresponding to the $(1, 0)$-form $d'\sigma$.*

Proof of Lemma. Since $d'\sigma = \sum \nabla_\alpha \sigma \, dz^\alpha$ and $S = \sum \nabla^{\bar{\alpha}} \sigma \dfrac{\partial}{\partial \bar{z}^\alpha}$, we obtain

$$d'' \circ \iota(d'\sigma) = d''\left(\sum \nabla^{\bar{\alpha}}\sigma \cdot \zeta_{\bar{\alpha}}\right) = \sum \left(\nabla_{\bar{\beta}} \nabla^{\bar{\alpha}} \zeta_{\bar{\alpha}} + \nabla^{\bar{\alpha}} \sigma \cdot \nabla_{\bar{\beta}} \zeta_{\bar{\alpha}}\right) d\bar{z}^\beta,$$

$$\iota(d'\sigma) \circ d'' = \sum \nabla^{\bar{\alpha}} \sigma (\nabla_{\bar{\alpha}} \zeta_{\bar{\beta}} - \nabla_{\bar{\beta}} \zeta_{\bar{\alpha}}) \, d\bar{z}^\beta.$$

Hence,

$$\begin{aligned}\Delta''_\sigma \zeta &= \Delta''\zeta - (d''\circ\iota(d'\sigma) + \iota(d'\sigma)\circ d'')\,\zeta \\ &= \Delta''\zeta - \sum (\nabla^{\bar\alpha}\sigma\cdot\nabla_\alpha\zeta_\beta + \nabla_\beta\nabla^{\bar\alpha}\sigma\cdot\zeta_{\bar\alpha})\,d\bar z^\beta \\ &= \Delta''\zeta - \nabla_S\zeta - \sum \nabla_\beta\nabla_\alpha\sigma\cdot\zeta^\alpha\,d\bar z^\beta .\end{aligned}$$

Now the lemma follows from the following fact:

$$Q_\sigma(\zeta) = Q(\zeta) - \sum \nabla_\beta\nabla_\alpha\sigma\,\zeta^\alpha\,d\bar z^\beta .$$

Suppose Z is holomorphic. Since S is of type (0, 1), Proposition 4.1 implies $\nabla_S\zeta = 0$. Then by Lemma,

$$\Delta''_\sigma\zeta - Q_\sigma(\zeta) = \Delta''\zeta - Q(\zeta).$$

The first half of Theorem 7.2 follows from Theorem 4.2.

To prove the second half, we make use of the following integral formula (see Appendix 2):

$$(\Delta''\zeta - Q(\zeta), \bar\omega) = (\nabla''\zeta, \overline{\nabla''\omega}),$$

where ζ and ω are arbitrary (0, 1)-forms. We set $\omega = e^\sigma\zeta$. Then the left hand side is equal to $(\Delta''\zeta - Q(\zeta), \bar\zeta)_\sigma$ while the right hand side is equal to

$$\begin{aligned}\int_M (\sum \nabla_\beta\zeta^\alpha\,\nabla^\beta(e^\sigma\zeta_{\bar\alpha}))\,dV &= \int_M (\sum \nabla_\beta\zeta^\alpha\cdot\nabla^\beta\sigma\cdot\zeta_{\bar\alpha} + \sum \nabla_\beta\zeta^\alpha\cdot\nabla^\beta\zeta_\alpha)\,e^\sigma\cdot dV \\ &= (\nabla_S\zeta, \bar\zeta)_\sigma + (\nabla''\zeta, \overline{\nabla''\zeta})_\sigma .\end{aligned}$$

Hence,

$$(\Delta''\zeta - Q(\zeta) - \nabla_S\zeta, \bar\zeta)_\sigma = (\nabla''\zeta, \overline{\nabla''\zeta})_\sigma .$$

By Lemma, this formula may be written as

$$(\Delta''_\sigma\zeta - Q_\sigma(\zeta), \bar\zeta)_\sigma = (\nabla''\zeta, \overline{\nabla''\zeta})_\sigma .$$

Now, if the left hand side vanishes, then $\nabla''\zeta = 0$. By Proposition 4.1, Z is holomorphic. q.e.d.

Let dv denote the volume element of a Kähler manifold M and define a positive $2n$-form $\Omega = e^\sigma\,dv$. Let $\mathfrak{h}_\sigma$ be the Lie algebra of holomorphic vector fields leaving the form Ω invariant, i.e.,

$$\mathfrak{h}_\sigma = \{Z \in \mathfrak{h}(M);\, L_Z\Omega = 0\}.$$

Proposition 7.3. *For a compact Kähler manifold M, the subalgebra $\mathfrak{h}_\sigma$ of the Lie algebra $\mathfrak{h}(M)$ of holomorphic vector fields may be defined by*

$$\begin{aligned}\mathfrak{h}_\sigma &= \{Z \in \mathfrak{h}(M);\, \delta''_\sigma\zeta = 0\} \\ &= \{Z \in \mathfrak{h}(M);\, \Delta''_\sigma\zeta = 0\},\end{aligned}$$

where ζ denotes the (0, 1)-form corresponding to Z.

Proof. We have

$$L_Z\Omega = L_Z(e^\sigma)\,dv + e^\sigma \cdot L_Z dv = L_Z\sigma \cdot \Omega - e^\sigma \cdot \delta''\zeta \cdot dv$$
$$= \big(\iota(d'\sigma)\zeta - \delta''\zeta\big)\Omega = -\delta''_\sigma\zeta \cdot \Omega .$$

This shows that $L_Z\Omega = 0$ if and only if $\delta''_\sigma\zeta = 0$.

By Proposition 4.1, $d''\zeta = 0$ whenever Z is holomorphic. Hence, for an element Z of $\mathfrak{h}(M)$, $\delta''_\sigma\zeta = 0$ is equivalent to $\delta''_\sigma\zeta = d''\zeta = 0$. This in turn is equivalent to $\Delta''_\sigma\zeta = 0$. q.e.d.

Theorem 7.4. *Let M be a compact Kähler manifold. Then the subalgebra $\mathfrak{h}_\sigma$ of $\mathfrak{h}(M)$ possesses the following properties:*

(1) *$\mathfrak{h}_\sigma$ is abelian;*

(2) *if $Z \in \mathfrak{h}_\sigma$ and $\omega(Z) = 0$ for all holomorphic 1-forms ω, then $Z = 0$ (that is, $\mathfrak{h}_\sigma \cap \mathfrak{h}_1 = 0$, where $\mathfrak{h}_1$ is the ideal of $\mathfrak{h}(M)$ introduced in § 3);*

(3) *$\dim \mathfrak{h}_\sigma \leq \frac{1}{2} b_1$, where b_1 is the first Betti number of M;*

(4) *if Z is a nonzero element of $\mathfrak{h}_\sigma$, then the zero set of Z is empty.*

Proof. We recall that $\mathfrak{h}_1$ is the ideal of $\mathfrak{h}(M)$ consisting of vector fields Z such that $\omega(Z) = 0$ for all holomorphic 1-forms ω. (Since M is a Kähler manifold, every holomorphic form is closed.) Since $\mathfrak{h}_1$ contains the derived algebra of $\mathfrak{h}(M)$ (see Proposition 3.1) and since $\mathfrak{h}_\sigma$ is a subalgebra, we have

$$[\mathfrak{h}_\sigma, \mathfrak{h}_\sigma] \subset \mathfrak{h}_\sigma \cap \mathfrak{h}_1 .$$

Let $Z \in \mathfrak{h}_\sigma \cap \mathfrak{h}_1$. By (2) of Theorem 4.4, we have $\zeta = d''f$. Hence,

$$(\zeta, \bar\zeta)_\sigma = (d''f, \bar\zeta)_\sigma = (f, \overline{\delta''_\sigma\zeta})_\sigma = 0$$

by Proposition 7.3. Therefore, $Z = 0$, thus proving (2) and hence (1). Now (3) follows from Proposition 3.1 and the inclusion

$$\mathfrak{h}_\sigma = \mathfrak{h}_\sigma/(\mathfrak{h}_\sigma \cap \mathfrak{h}_1) \subset \mathfrak{h}(M)/\mathfrak{h}_1 .$$

Suppose $Z \in \mathfrak{h}_\sigma$ and $\mathrm{Zero}(Z) \neq \emptyset$. As we remarked at the end of § 3, Z must be in $\mathfrak{h}_1$. By (2), $Z = 0$. This proves (4). q.e.d.

8. Compact Kähler Manifolds with Nonpositive First Chern Class

Let M be a compact complex manifold and $c_1(M)$ its first Chern class. We say that $c_1(M)$ is *nonpositive* and write $c_1(M) \leq 0$ if it can be represented by a closed (1, 1)-form

$$\gamma = \frac{i}{2\pi} \sum C_{\alpha\bar\beta}\, dz^\alpha \wedge d\bar z^\beta$$

such that $(C_{\alpha\bar{\beta}})$ is negative semi-definite hermitian. We have already considered manifolds with negative first Chern class (see § 2).

Assume that M is a compact Kähler manifold with Ricci tensor $R_{\alpha\bar{\beta}}$. Then $c_1(M)$ is represented by

$$\rho=\frac{i}{2\pi}\sum R_{\alpha\bar{\beta}}\,dz^\alpha\wedge d\bar{z}^\beta.$$

By Theorem 1 of Appendix 4, a (1, 1)-form ρ is cohomologous to γ if and only if there exists a real valued function σ such that

$$C_{\alpha\bar{\beta}}=R_{\alpha\bar{\beta}}-\nabla_{\bar{\beta}}\nabla_\alpha\sigma.$$

We prove two theorems of Lichnerowicz [6] (see also Matsushima [5]).

Theorem 8.1. *Let M be a compact Kähler manifold with $c_1(M)\leqq 0$ and $\mathfrak{h}(M)$ be the Lie algebra of holomorphic vector fields on M. Then*

(1) *$\mathfrak{h}(M)$ is abelian and* $\dim\mathfrak{h}(M)\leqq\frac{1}{2}b_1$.

(2) *If Z is a nonzero element of $\mathfrak{h}(M)$, it never vanishes on M.*

(3) *If a closed (1, 1)-form $\gamma=\frac{i}{2\pi}\sum C_{\alpha\bar{\beta}}\,dz^\alpha\wedge d\bar{z}^\beta$ represents $c_1(M)$ and if $(C_{\alpha\bar{\beta}})$ is negative semi-definite everywhere and negative definite somewhere, then $\mathfrak{h}(M)=0$.*

Proof. Let Z be a complex vector field of type (1, 0) with the corresponding (0, 1)-form ζ. By Theorem 7.2, Z is holomorphic if and only if

$$(\Delta''_\sigma\zeta-Q_\sigma(\zeta),\bar{\zeta})_\sigma=0.$$

This is equivalent to

$$(d''\zeta,\overline{d''\zeta})_\sigma+(\delta''_\sigma\zeta,\overline{\delta''_\sigma\zeta})-\sum C_{\alpha\bar{\beta}}\zeta^\alpha\overline{\zeta^\beta}=0.$$

Since $(C_{\alpha\bar{\beta}})$ is negative semi-definite, this equation is equivalent to

$$d''\zeta=0,\qquad \delta''_\sigma\zeta=0,\qquad \sum C_{\alpha\bar{\beta}}\zeta^\alpha\overline{\zeta^\beta}=0.$$

In other words, Z is holomorphic if and only if ζ is Δ''_σ-harmonic and satisfies $\sum C_{\alpha\bar{\beta}}\zeta^\alpha\overline{\zeta^\beta}=0$. By Theorem 7.1,

$$\dim\mathfrak{h}(M)\leqq\dim H^{0,1}(M;\mathbf{C}).$$

If $(C_{\alpha\bar{\beta}})$ is negative definite at a point p and Z is holomorphic, then ζ must vanish in a neighborhood of p and hence everywhere on M.

Let $\mathfrak{h}_1$ be the ideal of $\mathfrak{h}(M)$ consisting of holomorphic vector fields Z such that $\alpha(Z)=0$ for all holomorphic 1-forms α. In general, $\mathfrak{h}(M)/\mathfrak{h}_1$ is abelian (Proposition 3.1). We shall show that $\mathfrak{h}_1=0$. Let $Z\in\mathfrak{h}_1$. By

Theorem 4.4, $\zeta = d''f$, where f is a function. On the other hand, ζ is Δ''_σ-harmonic. Hence, $\zeta = 0$.

Since $\mathfrak{h}_1 = 0$, $\mathfrak{h}(M)$ itself is abelian. Suppose the zero set of Z is non-empty. If Z is holomorphic, for every holomorphic 1-form α the function $\alpha(Z)$ is holomorphic and hence constant. Since $\alpha(Z)$ vanishes at some point, it must vanish identically. This means that Z is an element of $\mathfrak{h}_1$. Hence, $Z = 0$. q.e.d.

Corollary 8.2. *Let M be a compact Kähler manifold with $c_1(M) \leqq 0$. If r is the maximal rank of $(C_{\alpha\bar{\beta}})$, then* $\dim \mathfrak{h}(M) \leqq n - r$, *where n is the complex dimension of M.*

Remark. Theorem 8.1 implies that if $c_1(M) < 0$, then $\mathfrak{h}(M) = 0$. But this is also a direct consequence of the vanishing theorem of Kodaira [1]-Nakano [1]. In fact, if we denote by Ω^p the sheaf of germs of holomorphic p-forms and by T and T^* the bundles of complex vectors of type $(1,0)$ and $(1,0)$-forms respectively, then

$$\mathfrak{h}(M) = H^0(M; \Omega^0(T)) \underset{\text{dual}}{\sim} H^n(M; \Omega^n(T^*)) = H^n(M; \Omega^1(K)),$$

where K is the canonical line bundle of M. If $c_1(M) < 0$, i.e., $c_1(K) > 0$, then the vanishing theorem implies $H^n(M; \Omega^1(K)) = 0$. But in Theorem 2.1, we proved a stronger statement that if $c_1(M) < 0$, then the group $\mathfrak{H}(M)$ of holomorphic transformations of M is finite.

Theorem 8.3. *Let M be a compact Kähler manifold with $c_1(M) = 0$. Let $\mathfrak{h}(M)$ be the Lie algebra of holomorphic vector fields of M. Let $\mathscr{C}^{1,0}$ denote the space of holomorphic* 1*-forms on M. Then*

(1) *$\mathfrak{h}(M)$ is abelian and* $\dim \mathfrak{h}(M) = \frac{1}{2} b_1$, *where b_1 is the first Betti number of M.*

(2) *The ideal $\mathfrak{h}_1 = \{Z \in \mathfrak{h}(M);\ \omega(Z) = 0$ for all $\omega \in \mathscr{C}^{1,0}\}$ is trivial. In particular, for every nonzero $Z \in \mathfrak{h}(M)$, its zero set is empty.*

(3) *The bilinear mapping $(\omega, Z) \in \mathscr{C}^{1,0} \times \mathfrak{h}(M) \to \omega(Z) \in \mathbf{C}$ is a dual pairing between $\mathscr{C}^{1,0}$ and $\mathfrak{h}(M)$.*

Proof. We prove the following lemma first.

Lemma. *Let M be a compact Kähler manifold with $c_1(M) \geqq 0$. Choose a function σ in such a way that $(C_{\alpha\bar{\beta}}) \geqq 0$ (see § 7). Let $H^{0,1}$ be the space of Δ''_σ-harmonic $(0,1)$-forms and let $\mathfrak{h}_\sigma$ be the ideal of $\mathfrak{h}(M)$ defined by $\mathfrak{h}_\sigma = \{Z \in \mathfrak{h}(M);\ \Delta''_\sigma \zeta = 0\}$ (see Proposition 7.3). Then $Z \to \zeta$ gives an isomorphism from $\mathfrak{h}_\sigma$ onto $H^{0,1}$.*

Proof of Lemma. Let $\zeta \in H^{0,1}$. We make use of the following formula established in the course of the proof of Theorem 7.2:

$$(\Delta''_\sigma \zeta - Q_\sigma(\zeta), \bar{\zeta})_\sigma = (\nabla'' \zeta, \overline{\nabla'' \zeta})_\sigma.$$

Since $\Delta''_\sigma \zeta = 0$ and $(C_{\alpha\bar{\beta}}) \geqq 0$, the left hand side is non-positive. Hence, $\nabla'' \zeta = 0$. By Proposition 4.1, the corresponding vector field Z is holomorphic. This proves that $h_\sigma \to H^{0,1}$ is surjective and hence is bijective. This completes the proof of Lemma.

Lemma implies that if $c_1(M) \geqq 0$, then

$$\dim \mathfrak{h}_\sigma = \tfrac{1}{2} b_1.$$

If $c_1(M) = 0$, then Theorem 8.1 implies that $\mathfrak{h}(M)$ is abelian and

$$\dim \mathfrak{h}(M) \leqq \tfrac{1}{2} b_1.$$

Hence, $\mathfrak{h}(M) = \mathfrak{h}_\sigma$ and

$$\dim \mathfrak{h}(M) = \tfrac{1}{2} b_1.$$

This proves (1). Since $\mathfrak{h}(M) = \mathfrak{h}_\sigma$, (2) of Theorem 7.4 implies $\mathfrak{h}_1 = 0$. This proves (2). We know (see § 3) that the bilinear mapping

$$\mathscr{C}^{1,0} \times \mathfrak{h}(M) \to \mathbf{C}$$

induces a dual pairing between $\mathscr{C}^{1,0}/\mathscr{D}^{1,0}$ and $\mathfrak{h}(M)/\mathfrak{h}_1$ (see also § 3 for the definition of $\mathscr{D}^{1,0}$). Since $\mathfrak{h}_1 = 0$ and $\dim \mathfrak{h}(M) = \frac{1}{2} b_1 = \dim \mathscr{C}^{1,0}$, it follows that $\mathscr{D}^{1,0} = 0$. Hence, the bilinear mapping above is a dual pairing between $\mathscr{C}^{1,0}$ and $\mathfrak{h}(M)$. q.e.d.

For results which sharpen Theorem 8.3, see Matsushima [4, 6].

9. Projectively Induced Holomorphic Transformations

Let M be a compact complex manifold and $\mathfrak{H}(M)$ the group of holomorphic transformations of M. Given a subgroup $\mathfrak{G}$ of $\mathfrak{H}(M)$, we ask if there exists an imbedding of M into a complex projective space $P_N(\mathbf{C})$ such that $\mathfrak{G}$ is induced by (i.e., the restriction of) a group of projective linear transformations of $P^N(\mathbf{C})$. We begin with the simplest case.

Theorem 9.1. *Let M be a compact complex manifold with positive or negative first Chern class $c_1(M)$. Then there is an imbedding of M into a complex projective space $P_N(\mathbf{C})$ such that every holomorphic transformation of M is induced by a unique projective linear transformation of $P_N(\mathbf{C})$.*

Proof. The case $c_1(M) < 0$ is contained in the proof of Theorem 2.1. Let K be the canonical line bundle of M. Then K is ample. For a sufficiently large integer p, the space $H^0(M; K^p)$ of holomorphic sections of K^p contains sufficiently many sections to induce an imbedding of M into $P_N(\mathbf{C})$, where $N = \dim H^0(M; K^p) - 1$. Every holomorphic transformation of M induces a linear transformation of $H^0(M; K^p)$ and hence a projective linear transformation of $P_N(\mathbf{C})$. See the proof of Theorem 2.1 for the details. If $c_1(M) > 0$, then K^{-1} is ample and a similar argument using $H^0(M; K^{-p})$ with a large integer p proves the theorem. q.e.d.

The following theorem is due to Blanchard [1] and Borel.

Theorem 9.2. *Let M be a compact Hodge manifold with first Betti number $b_1=0$. Then there is an imbedding of M into a complex projective space $P_N(\mathbf{C})$ such that every element of the largest connected group $\mathfrak{H}^0(M)$ of holomorphic transformations is induced by a unique projective linear transformation of M.*

Proof. Let Ω (resp. Ω^*) be the sheaf of germs of holomorphic functions (resp. holomorphic functions without zero) on M. The exact sequence

$$0 \to \mathbf{Z} \to \Omega \xrightarrow{\exp 2\pi \cdot} \Omega^* \to 1$$

induces an exact sequence

$$H^1(M;\Omega) \to H^1(M;\Omega^*) \to H^2(M;\mathbf{Z}).$$

It is essentially by definition that $H^1(M;\Omega^*)$ is the group of (isomorphism classes) of complex line bundles over M. By the theorem of Dolbeault, $H^1(M;\Omega)$ is isomorphic to the space $H^{0,1}$ of antiholomorphic (0, 1)-forms, i.e., harmonic (0, 1)-forms on M in a natural manner. Since $b_1=0$, we have $H^1(M;\Omega)=0$. Hence, $H^1(M;\Omega^*) \to H^2(M;\mathbf{Z})$ is injective. In other words, every complex line bundle of M (i.e., every element of $H^1(M;\Omega^*)$) is uniquely determined by its characteristic class (i.e., its image in $H^2(M;\mathbf{Z})$). Since $\mathfrak{H}^0(M)$ is a connected group, it acts trivially on $H^2(M;\mathbf{Z})$ and hence also on $H^1(M;\Omega^*)$. This means that if $f \in \mathfrak{H}^0(M)$ and L is a complex line bundle over M, then $f^*L=L$. In other words, f induces an automorphism $\tilde{f}$ of L compatible with f. It should be perhaps pointed out that the group $\mathfrak{H}^0(M)$ may not act on L since $f \circ g$ with $f, g \in \mathfrak{H}^0(M)$ may not induce $\tilde{f} \circ \tilde{g}$, that is, we may not have $(\widetilde{f \circ g}) = \tilde{f} \circ \tilde{g}$. This is due to the fact that f does not determine $\tilde{f}$ uniquely. If $\tilde{f}$ and $\tilde{\tilde{f}}$ are two automorphisms of L compatible with f, then $\tilde{\tilde{f}}(w) = \tilde{f}(w) \cdot a(z)$ for $w \in L$ and $z = \pi(w) \in M$, π denoting the projection $L \to M$. Since $a(z)$ is holomorphic in z and does not vanish, it is a nonzero constant. This shows that although $\tilde{f}$ is not unique, it is unique up to a nonzero constant multiple. Hence, $(\widetilde{f \circ g}) = (\tilde{f} \circ \tilde{g})\, a$, where a is a nonzero constant.

Since M is a Hodge manifold, the classical theorem of Kodaira [2] implies that there exists a very ample line bundle L so that $H^0(M;L)$ contains sufficiently many sections to induce an imbedding of M into $P_N(\mathbf{C})$, where $N = \dim H^0(M;L) - 1$. Each element f of $\mathfrak{H}^0(M)$ induces an automorphism $\tilde{f}$ of L, which in turn induces a linear transformation $\rho(f)$ of $H^0(M;L)$. Again, ρ may not be a representation in the strict sense but it satisfies the relation

$$\rho(f \circ g) = \rho(f) \circ \rho(g) \cdot a,$$

where a is a nonzero scalar. Hence, considered as a transformation of $P_N(\mathbf{C})$, $\rho(f)$ is uniquely determined by f and $\rho(f \circ g) = \rho(f) \circ \rho(g)$. q.e.d.

Remark 1. Actually we have shown in Theorem 9.2 that for any imbedding of M in $P_N(\mathbf{C})$ the group $\mathfrak{H}^0(M)$ can be realized as a transformation group of $P_N(\mathbf{C})$. If we choose an imbedding of M into $P_N(\mathbf{C})$ more carefully, the group $\mathfrak{H}^0(M)$ can be given by a linear group acting on $\mathbf{C}^{N+1}$ $(=H^0(M; L))$, that is, ρ can be made into a representation of $\mathfrak{H}^0(M)$ on $H^0(M; L)$ in the strict sense. This follows from the following general result.

Proposition 9.3. *There is an imbedding of* $P_N(\mathbf{C})$ *into some* $P_m(\mathbf{C})$ *such that the projective linear group* $\mathrm{PGL}(N; \mathbf{C})$ *is represented by linear transformations of the corresponding vector space* $\mathbf{C}^{m+1}$.

Proof. Let V be the space of symmetric tensors over $\mathbf{C}^{N+1}$ of degree $N+1$, i.e., the space of homogeneous polynomials of degree $n+1$ over the dual space of $\mathbf{C}^{N+1}$. We consider $\mathrm{PGL}(N; \mathbf{C})$ as the image of $\mathrm{SL}(N+1; \mathbf{C})$ acting on $\mathbf{C}^{N+1}$. Then the center $\{a I_{N+1}; a^{N+1} = 1\}$ of $\mathrm{SL}(N+1; \mathbf{C})$ is precisely the set of elements which induce the trivial transformation of $P_N(\mathbf{C})$. The group $\mathrm{SL}(N+1; \mathbf{C})$ acts on V in the obvious manner and its center acts trivially on V. Hence $\mathrm{PGL}(N; \mathbf{C})$ has a faithful representation in $\mathrm{GL}(V)$. This is compatible with the obvious imbedding of $P_N(\mathbf{C})$ into $P_m(\mathbf{C})$, where $m+1 = \dim V$. q.e.d.

Remark 2. In Theorem 9.2, if $H^2(M; \mathbf{Z}) = \mathbf{Z}$, then not only the identity component $\mathfrak{H}^0(M)$ but also the whole group $\mathfrak{H}(M)$ acts trivially on $H^2(M; \mathbf{Z})$ and hence $\mathfrak{H}(M)$ can be realized as a group of projective linear transformations compatible with a projective imbedding of M.

For complex tori, we cannot have results similar to Theorems 9.1 and 9.2. In fact, we prove the following theorem of Blanchard [1].

Theorem 9.4. *Let* M *be a Hodge manifold,* $\mathfrak{H}(M)$ *the group of holomorphic transformations and* $\mathfrak{h}(M)$ *the Lie algebra of holomorphic vector fields of* M. *As in* § 3, *let* $\mathfrak{h}_1$ *be the ideal of* $\mathfrak{h}(M)$ *consisting of* $Z \in \mathfrak{h}(M)$ *such that* $\alpha(Z) = 0$ *for all holomorphic* 1*-forms* α. *Let* $\mathfrak{H}_1$ *be the connected normal subgroup of* $\mathfrak{H}(M)$ *generated by* $\mathfrak{h}_1$. *Then given a connected Lie subgroup* $\mathfrak{G}$ *of* $\mathfrak{H}(M)$, *there is an imbedding of* M *into* $P_N(\mathbf{C})$ *such that every element of* $\mathfrak{G}$ *is induced by a unique projective linear transformation of* $P_N(\mathbf{C})$ *if and only if* $\mathfrak{G}$ *is a subgroup of* $\mathfrak{H}_1$.

The proof is divided into several lemmas.

Lemma 1. *Let* M *be a complex submanifold of a Kähler manifold* P. *Let* Z *be a holomorphic vector field on* P *and* ζ *the corresponding* $(0,1)$*-form on* P. *If the restriction of* Z *to* M *is a (holomorphic) tangent vector field of* M, *i.e.,* Z *is tangent to* M *at each point of* M, *then the restriction of* ζ *to* M *is the* $(0,1)$*-form corresponding to the restriction of* Z *to* M.

Proof of Lemma 1. Let $(\ ,\ )_P$ and $(\ ,\ )_M$ denote the inner products defined by the Kähler metric of P and the induced Kähler metric of M, respectively. Let W be any complex tangent vector of M of type $(1,0)$. Then

$$(Z,\overline{W})_M=(Z,\overline{W})_P=\zeta(\overline{W}).$$

This proves Lemma 1.

Lemma 2. *Let M be a closed complex submanifold of a compact Kähler manifold P. Denote the ideal $\mathfrak{h}_1$ of $\mathfrak{h}(M)$ defined in Theorem* 9.4 *by $\mathfrak{h}_1(M)$ and denote the similarly defined ideal of $\mathfrak{h}(P)$ by $\mathfrak{h}_1(P)$. Let Z be an element of $\mathfrak{h}_1(P)$ such that its restriction to M is tangent to M. Then its restriction to M is in $\mathfrak{h}_1(M)$.*

Proof of Lemma 2. Let ζ be the (0, 1)-form corresponding to Z. By Theorem 4.4, $\zeta=d''f$ for some function f. Then $\zeta|_M=d''(f|_M)$. By Theorem 4.4 and Lemma 1, $Z|_M$ is in $\mathfrak{h}_1(M)$.

We are now in a position to prove a half of Theorem 9.4. Let P be a compact Kähler manifold without (non-trivial) holomorphic 1-form such as a complex projective space. Then $\mathfrak{h}_1(P)=\mathfrak{h}(P)$. Then Lemma 2 implies that any holomorphic vector field on M which comes from a holomorphic vector field of P is in the ideal $\mathfrak{h}_1(M)$. This shows that in Theorem 9.4 if $\mathfrak{G}$ comes from projective linear transformations of $P_N(\mathbf{C})$ in which M is imbedded, then $\mathfrak{G}$ is contained in $\mathfrak{H}_1$.

To prove the remaining half of the theorem, it is clear from the proof of Theorem 9.2 that all we have to do is to find a very ample line bundle L over M such that every transformation belonging to $\mathfrak{H}_1$ can be lifted to an automorphism of L. In other words, it suffices to find a very ample line bundle L over M such that every holomorphic vector field of M belonging to $\mathfrak{h}_1$ can be lifted to a holomorphic vector field on L. Actually we prove

Lemma 3. *Let M be a Hodge manifold and L an ample line bundle over M. Then every holomorphic vector field of M belonging to $\mathfrak{h}_1$ can be lifted to a holomorphic vector field of L.*

Proof of Lemma 3. Choosing a hermitian fibre metric h in the line bundle L, let ω be its connection form on the associated principal bundle L^*. We note that L^* can be obtained from L by deleting the zero section of L. Since L is ample, we may assume that h was so chosen that its curvature is positive. We shall write down h, ω and the curvature explicitly.

Let U be a small open set in M with local coordinate system $z^1,\dots,z^n$. We fix a holomorphic section σ of L^* over U. Then we can identify $L^*|_U$ with $U\times\mathbf{C}^*$. We denote by t the natural coordinate system in $\mathbf{C}^*$. Then we can take $z^1,\dots,z^n,t$ as a coordinate system in $L^*|_U=U\times\mathbf{C}^*$.

Let h_U be the square of the length of the section σ; it is a positive function on U. We define a (1,0)-form ω_U on U by

$$\omega_U = d'(\log h_U).$$

Then ω_U is the pull-back of the connection form ω by σ, and we have

$$\omega = \omega_U + \frac{1}{t}\, dt = d'(\log h_U\, t\, \bar{t}).$$

The curvature form is given by

$$d''\omega_U = d''d'(\log h_U) = \sum g_{i\bar{k}}\, dz^i \wedge d\bar{z}^k,$$

where

$$g_{i\bar{k}} = -\partial^2 \log h_U / \partial z^i\, \partial \bar{z}^k.$$

We use $2\sum g_{i\bar{k}}\, dz^i\, d\bar{z}^k$ as our Kähler metric on M. Let

$$Z = \sum \zeta^j \frac{\partial}{\partial z^j}$$

be a holomorphic vector field on M belonging to $\mathfrak{h}_1$ and ζ the corresponding (0,1)-form. By Theorem 4.4, there is a function f on M such that

$$\zeta = d'' f.$$

We lift Z to a vector field $\tilde{Z}$ on L^* of type (1,0) such that

$$-f = \omega(\tilde{Z}).$$

This determines $\tilde{Z}$ uniquely. If we write

$$\tilde{Z} = \sum \zeta^j \frac{\partial}{\partial z^j} + \tau \frac{\partial}{\partial t},$$

then

$$-f = \omega(\tilde{Z}) = \omega_U(Z) + \frac{\tau}{t}.$$

We want to show that $\tilde{Z}$ is holomorphic, i.e., τ is holomorphic. We have

$$-d''f = d''(\omega_U(Z)) + d''(\tau/t) = d'' \circ \iota_Z \omega_U + d''(\tau/t).$$

Since Z is holomorphic, we have

$$d'' \circ \iota_Z + \iota_Z \circ d'' = 0.$$

Hence,

$$\begin{aligned}-d''f &= -\iota_Z \circ d''\omega_U + d''(\tau/t)\\ &= -\iota_Z\left(\sum g_{j\bar{k}}\, dz^j \wedge d\bar{z}^k\right) + d''(\tau/t)\\ &= -\sum g_{j\bar{k}}\, \zeta^j\, d\bar{z}^k + d''(\tau/t) = -\zeta + d''(\tau/t).\end{aligned}$$

Since $d''f=\zeta$, it follows that $d''(\tau/t)=0$. Since $d''t=0$, this implies $d''\tau=0$, thus proving that τ is holomorphic. We have thus proved that Z can be lifted to a holomorphic vector field $\tilde{Z}$ on L^*. Hence, Z can be also lifted to a holomorphic vector field on the associated bundle L. This completes the proof of Lemma 3. q.e.d.

From Lemma 3 in the proof above, we obtain

Proposition 9.5. *Let M and $\mathfrak{h}_1$ be as in Theorem 9.4 and L be a complex line bundle over M. Then every vector field on M belonging to $\mathfrak{h}_1$ can be lifted to a holomorphic vector field on L.*

Proof. We have proved this proposition when L is ample. It suffices therefore to show that the given line bundle L is of the form $L_1 L_2^{-1}$, where both L_1 and L_2 are ample line bundles. Take any ample line bundle F over M and let k be a large integer such that LF^k is ample. Then set $L_1=LF^k$ and $L_2=F^k$. q.e.d.

From the definition of $\mathfrak{h}_1$ it is clear that a holomorphic vector field Z on M belongs to $\mathfrak{h}_1$ if its zero set is non-empty. It follows that if the Euler number of M is nonzero, then every holomorphic vector field Z belongs to $\mathfrak{h}_1$. More generally (see § 12, Corollary 12.2), if some Chern number of M is nonzero, then every holomorphic vector field Z vanishes at some point and hence belongs to $\mathfrak{h}_1$. Hence

Theorem 9.6. *Let M be a Hodge manifold such that some Chern number (e.g., the Euler number) is nonzero. Then M can be imbedded into $P_N(\mathbf{C})$ in such a way that the largest connected group $\mathfrak{H}^0(M)$ of holomorphic transformations is induced by projective linear transformations of $P_N(\mathbf{C})$.*

We shall now prove the following theorem of Matsushima [4].

Theorem 9.7. *Let M be a Hodge manifold and $\mathfrak{h}_1$ the Lie algebra of holomorphic vector fields Z such that $\alpha(Z)=0$ for all holomorphic 1-forms α. Then $\mathfrak{h}_1$ coincides with the set of holomorphic vector fields with non-empty zero set.*

Proof. We know that a holomorphic vector field Z with non-empty zero set belongs to $\mathfrak{h}_1$ because $\alpha(Z)$ is constant for any holomorphic 1-form α. Conversely, if Z belongs to $\mathfrak{h}_1$, then Z is induced by a holomorphic vector field of a projective space $P_N(\mathbf{C})$ in which M is imbedded. It suffices therefore to prove the following

Lemma. *Let M be a closed complex subspace in $P_N(\mathbf{C})$ and φ_t be a 1-parameter group of projective linear transformations of $P_N(\mathbf{C})$ leaving M invariant. Then φ_t has a fixed point in M.*

Proof of Lemma. The proof is by induction on N. Lemma is trivially true for $N=1$. Assume that Lemma holds for $N-1$. Represent φ_t as a 1-parameter group of linear transformations of the vector space $\mathbf{C}^{N+1}$. If we choose a suitable basis for $\mathbf{C}^{N+1}$, we can represent φ_t by triangular matrices with zeros below the main diagonal. Let $e_0, e_1, \ldots, e_N$ be such a basis and V be the N-dimensional vector subspace of $\mathbf{C}^{N+1}$ spanned by $e_0, \ldots, e_{N-1}$. Then V is invariant by φ_t. If we denote the projective space associated to V by $P_{N-1}(\mathbf{C})$, then $P_{N-1}(\mathbf{C})$ is a hyperplane in $P_N(\mathbf{C})$ invariant by φ_t. The hyperplane section $P_{N-1}(\mathbf{C}) \cap M$ is a (non-empty) closed complex subspace of $P_{N-1}(\mathbf{C})$ invariant by φ_t. By the inductive assumption, φ_t has a fixed point in $P_{N-1}(\mathbf{C}) \cap M$. q.e.d.

As a consequence of Theorem 9.7, we obtain a result of Matsushima [4].

Theorem 9.8. *Let M be a Hodge manifold with a holomorphic vector field Z with empty zero set. Then M admits a holomorphic 1-form α such that $\alpha(Z) \neq 0$. In particular, the first Betti number b_1 of M is positive.*

Proof. By Theorem 9.7, Z is not in the ideal $\mathfrak{h}_1$ so that $\alpha(Z) \neq 0$ for some holomorphic 1-form α. q.e.d.

Theorem 9.4 is weaker than the result proved by Blanchard [1]. For the original theorem of Blanchard, see also Matsushima [5].

10. Zeros of Infinitesimal Isometries

Let M be an n-dimensional compact Kähler manifold and $Z = X - iJX$ be a holomorphic vector field such that its real part X is an infinitesimal isometry. Assume that the zero set $\mathrm{Zero}(Z)$ $(=\mathrm{Zero}(X))$ is nonempty. This is equivalent to assuming that $J\xi = du$, where ξ is the 1-form corresponding to X and u is a real valued function on M (see Corollary 4.6). We shall apply the Morse theory to the function u. A point of M is a critical point of u if and only if it is a zero point of X. Let $\mathrm{Zero}(X) = \bigcup N_i$ be the decomposition of $\mathrm{Zero}(X)$ into its connected components. Then each N_i is a closed totally geodesic submanifold of even co-dimension by Theorem 5.3 of Chapter II. Since X preserves the complex structure J, its covariant derivative ∇X commutes with J. If $x \in N_i$, then for a suitable orthonormal basis of $T_x(M)$ the linear endomorphisms $(\nabla X)_x$ and J_x of $T_x(M)$ are given by matrices of the form (see the proof of Theorem 5.3, Chapter II):

$$(\nabla X)_x = \begin{pmatrix} 0 & & & & & & & \\ & \ddots & & & & & & \\ & & 0 & & & & & \\ & & & 0 & a_1 & & & \\ & & & -a_1 & 0 & & & \\ & & & & & \ddots & & \\ & & & & & & 0 & a_k \\ & & & & & & -a_k & 0 \end{pmatrix}, \quad J_x = \begin{pmatrix} 0 & 1 & & & \\ -1 & 0 & & & \\ & & \ddots & & \\ & & & 0 & 1 \\ & & & -1 & 0 \end{pmatrix},$$

where $a_i \neq 0$ for $i = 1, \ldots, k$. The subspace of $T_x(M)$ spanned by the first $2n-2k$ elements (resp. the last $2k$ elements) of the basis is the tangent space $T_x(N_i)$ (resp. the normal space $T_x^{\perp}(N_i)$ to N_i). It follows that $T_x(N_i)$ is invariant by J so that N_i is a complex submanifold of M. Since $du = J\xi$, the Hessian matrix of u at x is given by $J_x \circ (\nabla X)_x$. The latter is a diagonal matrix with diagonal elements $0, \ldots, 0, -a_1, -a_1, \ldots, -a_k, -a_k$. The nullity of the Hessian of u at x is equal to $2n-2k$ and hence N_i is a non-degenerate critical submanifold. The index of the Hessian of u at x, i.e., the number of the negative diagonal entries in $J_x \circ (\nabla X)_x$, is clearly even since the nonzero diagonal elements appear in pairs.

We shall now sketch an outline of the proof of the following theorem of Frankel [1].

Theorem 10.1. *Let X be an infinitesimal isometry on a compact Kähler manifold M with nonempty zero set* $\mathrm{Zero}(X)$. *Then for any coefficient field K,*

$$\sum_q \dim H_q(M; K) = \sum_q \dim H_q(\mathrm{Zero}(X); K).$$

More precisely, if $\mathrm{Zero}(X) = \bigcup N_i$ *is the decomposition into connected components, then*

$$\dim H_q(M; K) = \sum_i \dim H_{q-\lambda_i}(N_i; K),$$

where λ_i is the number of the negative eigen-values of $J_x \circ (\nabla X)_x$ for $x \in N_i$ and hence is an even integer.

Proof. By Theorem 4.3, the complex vector field $Z = X - iJX$ is holomorphic. To the function u constructed above, we apply the following result of Bott [3]:

Lemma. *Let u be a real valued function on a manifold M such that $M_r = \{x \in M; u(x) \leqq r\}$ is compact for every real number r. Let $a < c < b$ be* *real numbers such that c is the only critical value of u in the interval* $[a, b]$ *and assume that* $N = u^{-1}(c)$ *is a non-degenerate critical submanifold. If* $N = \bigcup N_i$ *is the decomposition into connected components, then for any coefficient field K we have*

$$H_q(M_b, M_a; K) \cong \sum_i H_{q-\lambda_i}(N_i; K),$$

where λ_i *is the index of the Hessian of u at a point of N.*

The proof is similar to that of the case where N is a non-degenerate critical point (see Milnor [1]).

Let $c_1 c_2 \ldots$ be the critical values of u and $a_0, a_1, \ldots$ be real numbers such that $a_0 < c_1 < a_1 < c_2 < a_2 < \cdots$. Setting $M_p = M_{a_p} = \{x \in M; u(x) \leqq a_p\}$ for $p = 0, 1, 2, \ldots$, we filter the group $C(M)$ of singular chains by the subgroups $C(M_p)$. Then in the resulting spectral sequence, we have

$$E^1_{p,r} = H_{p+r}(M_p, M_{p-1}; K).$$

Applying Lemma 1 to the right hand side and writing $q-p$ for r, we obtain

$$\sum_p E^1_{p,q-p} = \sum_i H_{q-\lambda_i}(N_i; K).$$

From

$$\dim H_q(M; K) = \dim \sum_p E^\infty_{p,q-p} \leqq \dim \sum_p E^1_{p,q-p},$$

we obtain

$$\dim H_q(M; K) \leqq \sum_i \dim H_{q-\lambda_i}(N_i; K).$$

This establishes the inequality in one direction. The theorem follows from a result of Floyd [1] and Conner [1] (see Theorem 5.5, Chapter II), which gives the inequality in the other direction. q.e.d.

In Borel [3; p. 171], Theorem 10.1 is proved for a toral group acting on a homologically compact Kähler manifold, e.g., a compact symplectic manifold (with a coefficient field K of characteristic 0).

The following corollaries are also due to Frankel.

Corollary 10.2. *Let X be as in Theorem* 10.1 *with nonempty* Zero(X). *Then*

(1) Zero(X) *has torsion if and only if M has torsion.*

(2) $H_{2i+1}(\text{Zero}(X); \mathbf{Z}) = 0$ *for all i if and only if* $H_{2i+1}(M; \mathbf{Z}) = 0$ *for all i.*

Proof. (1) This follows from Theorem 10.1 and from the fact that a space has no torsion if and only if its rational Betti numbers coincide with its mod p Betti numbers for all primes p.

(2) Assume $H_{2i+1}(\text{Zero}(X); \mathbf{Z})=0$ for all i. Since each connected component N_i of $\text{Zero}(X)$ is a complex submanifold and hence an orientable manifold of even dimension, the Poincaré duality theorem implies that N_i has no torsion. By (1) M has no torsion. Since the indices λ_i in Theorem 10.1 are all even, Theorem 10.1 implies that $H_{2i+1}(M; \mathbf{Q})=0$ for all i. Hence, $H_{2i+1}(M; \mathbf{Z})=0$ for all i. The implication in the other direction can be proved in a similar fashion. q.e.d.

Corollary 10.3. *Let X be as in Theorem 10.1 with nonempty $\text{Zero}(X)$. If $\text{Zero}(X)$ is discrete, then M has no torsion and its odd dimensional Betti numbers vanish.*

Recently, E. Wright [1] improved Corollary 10 3 by proving the following result.

Theorem 10.4. *Let X be as in Theorem 10.1 with nonempty $\text{Zero}(X)$. If $\text{Zero}(X)$ is discrete, then*

$$H^{p,q}(M; \mathbf{C})=0 \quad \text{for } p \neq q.$$

11. Zeros of Holomorphic Vector Fields

In this section, we shall describe recent results of Howard [1] on holomorphic vector fields with nonempty zero set.

Theorem 11.1. *Let M be a compact Kähler manifold with a nontrivial holomorphic vector field Z such that $\text{Zero}(Z)$ is nonempty. Then*

$$H^{n,0}(M; \mathbf{C})=0 \quad (n=\dim M),$$

that is, M admits no nonzero holomorphic n-forms.

Proof. Let ζ be the (0, 1)-form corresponding to Z. Since $\text{Zero}(Z)$ is nonempty, there is a function f on M such that $\zeta=d''f$ (see Corollary 4.5). Since

$$Z(\bar{f})=d'\bar{f}(Z)=\bar{\zeta}(Z)=\|Z\|^2,$$

$Z(\bar{f})\geqq 0$ and >0 almost everywhere.

Let φ be a holomorphic n-form on M. Since every holomorphic form on M is necessarily closed, we obtain

$$L_Z\varphi=d\circ\iota_Z\varphi+\iota_Z\circ d\varphi=0.$$

Since $d\bar{\varphi}=(\overline{d\varphi})=0$ and $\iota_Z\bar{\varphi}=0$, we obtain

$$L_Z\bar{\varphi}=d\circ\iota_Z\bar{\varphi}+\iota_Z\circ d\bar{\varphi}=0.$$

Hence,

$$L_Z(\varphi\wedge\bar{\varphi})=0.$$

Making use of this formula, we obtain

$$Z(\bar{f})\cdot\varphi\wedge\bar{\varphi}=L_Z(\bar{f}\cdot\varphi\wedge\bar{\varphi})-\bar{f}L_Z(\varphi\wedge\bar{\varphi})=L_Z(\bar{f}\cdot\varphi\wedge\bar{\varphi}).$$

Since $\bar{f}\cdot\varphi\wedge\bar{\varphi}$ is a form of degree (n, n), it is closed and hence

$$\int_M L_Z(\bar{f}\varphi\wedge\bar{\varphi})=d\circ\iota_Z(\bar{f}\cdot\varphi\wedge\bar{\varphi})=0$$

by Stokes' theorem. Hence,

$$\int_M Z(\bar{f})(i^{n^2}\varphi\wedge\bar{\varphi})=0.$$

(The factor i^{n^2} makes the (n,n)-form $i^{n^2}\varphi\wedge\bar{\varphi}$ real and non-negative.) Since $Z(\bar{f})=\|Z\|^2$ is positive almost everywhere, we conclude that φ vanishes identically on M. q.e.d.

Since a holomorphic n-form is a holomorphic section of the canonical line bundle K, the following theorem generalizes that of Howard.

Theorem 11.2. *Let M be a compact Kähler manifold with a nontrivial holomorphic vector field Z such that* Zero(Z) *is nonempty. Let K be the canonical line bundle of M and p be a positive integer. Then the line bundle K^p admits no nonzero holomorphic sections.*

Proof. Let $\zeta=d''f$ be the $(0,1)$-form corresponding to Z as in the proof of Theorem 11.1 so that $Z(\bar{f})=\|Z\|^2$. Let φ be a holomorphic section of K^p. As in the proof of Theorem 2.1, in terms of a local coordinate system φ may be symbolically expressed as follows:

$$\varphi=h(dz^1\wedge\cdots\wedge dz^n)^p,$$

where h is a holomorphic function defined in the coordinate neighborhood. Define a real, non-negative (n,n)-form $|\varphi|^{2/p}$ by

$$|\varphi|^{2/p}=i^{n^2}|h|^{2/p}\,dz^1\wedge\cdots\wedge dz^n\wedge d\bar{z}^1\wedge\cdots\wedge d\bar{z}^n.$$

In the proof of Theorem 2.1 we established that the largest connected group of holomorphic transformations of M leaves every holomorphic section of K^p invariant. Hence $|\varphi|^{2/p}$ is invariant by the 1-parameter group generated by Z, that is, $L_Z(|\varphi|^{2/p})=0$. The remainder of the proof is essentially the same as the proof of Theorem 11.1. Since

$$\begin{aligned}Z(\bar{f})|\varphi|^{p/2}&=L_Z(\bar{f}|\varphi|^{p/2})-\bar{f}L_Z(|\varphi|^{p/2})=L_Z(\bar{f}|\varphi|^{p/2})\\&=d\circ\iota_Z(\bar{f}|\varphi|^{p/2}).\end{aligned}$$

Integrating the both ends of the equalities above and making use of Stokes' theorem, we obtain

$$\int_M Z(\bar{f})|\varphi|^{p/2}=0.$$

Since $Z(\bar{f}) = \|Z\|^2$ is positive almost everywhere, we conclude that $|\varphi|^{p/2}$ and hence φ vanish identically on M. q.e.d.

The following theorem of Howard sharpens Theorem 11.1 for Hodge manifolds.

Theorem 11.3. *Let M be a Hodge manifold with a nontrivial holomorphic vector field Z such that* Zero(Z) *is nonempty. Then M admits no nonzero holomorphic p-form for $p >$* dim Zero(Z).

The set Zero(Z) is a subvariety (possibly with singularities) of M, and by dim Zero(Z) we mean the maximum dimension of its components.

Proof. Let $n = \dim M$. By Theorem 9.4, there is an imbedding of M into a complex projective space $P_N(\mathbf{C})$ such that the holomorphic vector field Z can be extended to a holomorphic vector field, denoted also by Z, of $P_N(\mathbf{C})$. Since the automorphism group of $P_N(\mathbf{C})$ is finitely covered by the group $\mathrm{SL}(N+1; \mathbf{C})$, Z may be considered as an element of the Lie algebra of $\mathrm{SL}(N+1; \mathbf{C})$. With respect to a suitable basis of $\mathbf{C}^{N+1}$, Z is represented by a triangular matrix, and let $z^0, z^1, \ldots, z^N$ be the coordinate system in $\mathbf{C}^{N+1}$ with respect to such a basis. Let L_k be the k-dimensional linear subspace of $P_N(\mathbf{C})$ defined by

$$z^{k+1} = \cdots = z^N = 0.$$

Then Z is tangent to L_k at each point of L_k for $k = 0, 1, \ldots, N$. Let ω denote the Kähler (1,1)-form of $P_N(\mathbf{C})$ corresponding to an invariant Kähler metric. Multiplying ω by a suitable positive constant, we may assume that ω represents the characteristic class of the line bundle over $P_N(\mathbf{C})$ defined by a hyperplane. Thus ω represents the second cohomology class dual to the $2(n-1)$-dimensional homology class represented by a hyperplane.

Lemma 1. *Let M and Z be as above. If φ is a nonzero holomorphic p-form on M, then there exist r-dimensional irreducible subvarieties V_r of M for $r = n, n-1, \ldots, p$ with the following properties:*

(1) *Z is tangent to V_r at every regular point of V_r.*

(2) *If j_r: $V_r \to M$ denotes the imbedding and $[\varphi \wedge \omega^{r-p}]$ denotes the cohomology class in $H^{2r-p}(M; \mathbf{C})$ represented by the $(r, r-p)$-form $\varphi \wedge \omega^{r-p}$, then $j_r^* [\varphi \wedge \omega^{r-p}]$ is a nonzero element of $H^{2r-p}(V_r; \mathbf{C})$.*

The proof of Lemma 1 is by induction on r starting with $r = n$. For $r = n$, it suffices to take $V_n = M$. It is well known that the multiplication by ω^{n-p} defines an isomorphism of $H^p(M; \mathbf{C})$ onto $H^{2n-p}(M; \mathbf{C})$ (see for example Weil [1; p.75]). Hence $[\varphi \wedge \omega^{n-p}]$ is nonzero. Assuming Lemma 1 for r, let k be the smallest integer such that L_k contains V_r.

Viewing V_r as an algebraic variety in the k-dimensional projective space L_k, we consider the hyperplane section $V_r \cap L_{k-1}$. Since ω is the characteristic class of the line bundle defined by a hyperplane, we obtain

$$[\varphi \wedge \omega^{r-1-p}] \,|(V_r \cap L_{k-1}) = [\varphi \wedge \omega^{r-p}] |\, V_r.$$

The right hand side is nonzero by assumption. We write $V_r \cap L_{k-1} = \sum m_i D_i$, where D_i's are irreducible divisors of V_r and m_i's are integers. Since $[\varphi \wedge \omega^{r-1-p}]|(V_r \cap L_{k-1}) \neq 0$, it follows that $[\varphi \wedge \omega^{r-1-p}]|D_i \neq 0$ for some D_i. We denote such a divisor D_i by V_{r-1}. Then clearly V_{r-1} possesses the properties (1) and (2) for $r-1$. This completes the proof of Lemma 1.

Let V be the p-dimensional irreducible subvariety V_p obtained in Lemma 1 and V' be the set of regular points of V. Denote the inclusion map $V' \to M$ by j. Then as in the proof of Theorem 11.1, we have

$$\begin{aligned} Z(\bar{f}) \cdot j^*(\varphi \wedge \bar{\varphi}) &= j^* L_Z(\bar{f}\varphi \wedge \bar{\varphi}) = j^* d \circ \iota_Z(\bar{f}\varphi \wedge \bar{\varphi}) + j^* \iota_Z \circ d(\bar{f}\varphi \wedge \bar{\varphi}) \\ &= j^* d \circ \iota_Z(\bar{f}\varphi \wedge \bar{\varphi}). \end{aligned}$$

Now we make use of the following Stokes theorem (see P. Lelong [1; § 6]):

Lemma 2. *Let V be a p-dimensional subvariety of a compact complex manifold M and V' be the set of regular points of V. Let Φ be a $(2p-1)$-form defined in a neighborhood of V. Then*

$$\int_{V'} d\Phi = 0.$$

Hence,

$$\int_{V'} Z(\bar{f}) \cdot j^*(\varphi \wedge \bar{\varphi}) = \int_{V'} j^* d \circ \iota_Z(\bar{f}\varphi \wedge \bar{\varphi}) = 0.$$

As in the proof of Theorem 11.1, we conclude that $\varphi = 0$ since Z cannot vanish identically on V' by our assumption $\dim \mathrm{Zero}(Z) < p = \dim V'$.

Corollary 11.4. *Let M be a Hodge manifold with a holomorphic vector field Z whose zero set* Zero(Z) *is nonempty and discrete. Then*

(1) *M admits no nonzero holomorphic p-forms for $p > 0$;*

(2) *the arithmetic genus of M is equal to 1;*

(3) *the fundamental group $\pi_1(M)$ has no proper subgroup of finite index and $H_1(M; \mathbf{Z}) = 0$.*

Proof. (1) is immediate from Theorem 11.3. (2) follows also immediately since the arithmetic genus of M is equal to

$$\sum (-1)^p \dim H^{p,0}(M; \mathbf{C}) = \dim H^{0,0}(M; \mathbf{C}) = 1.$$

To prove (3), assume that $\pi_1(M)$ has a subgroup of finite index k and let $\tilde{M}$ be the corresponding k-sheeted covering space of M. Lift Z to a holomorphic vector field $\tilde{Z}$ of $\tilde{M}$ and apply (2) to $\tilde{M}$. Then the arithmetic genus of $\tilde{M}$ is also 1. On the other hand, the Riemann-Roch-Hirzebruch theorem implies that the arithmetic genus of $\tilde{M}$ is k times that of M. Hence, $k=1$. Since $H_1(M;\mathbf{Z})=\pi_1(M)/[\pi_1(M),\pi_1(M)]$ and $H_1(M;\mathbf{R})=0$, it follows that $[\pi_1(M),\pi_1(M)]$ is a subgroup of $\pi_1(M)$ of finite index and hence, $H_1(M;\mathbf{Z})=0$. q.e.d.

(3) of the corollary above is due to Lichnerowicz [8]. He has shown many other results on zeros of holomorphic vector fields on compact Kähler manifolds with non-negative first Chern class.

12. Holomorphic Vector Fields and Characteristic Numbers

Let M be an n-dimensional compact complex manifold and Z be a holomorphic vector field on M. Let $\mathrm{Zero}(Z)=\bigcup N_i$ be the decomposition of the zero set of Z into its connected components N_i. We assume that Z is non-degenerate along N_i in the following sense. Since we consider one N_i at a time, we denote N_i by N. First, we assume that N is a complex submanifold of codimension r so that $\dim N=n-r$. Let $T^{1,0}(M)$ and $T^{1,0}(N)$ be the holomorphic vector bundles of complex vectors of type (1,0) over M and N, respectively. We set

$$E=T^{1,0}(M)|N, \qquad E'=T^{1,0}(N).$$

Then E is a holomorphic vector bundle of rank n over N and E' is a holomorphic vector subbundle of E of rank $n-r$. The holomorphic vector field Z induces an endomorphism of the bundle E; in terms of a local coordinate system $z^1, \dots, z^n$ of M it is given by the matrix $(\partial\zeta^\alpha/\partial z^\beta)$ if $Z=\sum \zeta^\alpha \dfrac{\partial}{\partial z^\alpha}$. We denote this holomorphic endomorphism of E by A. (Since Z vanishes on N, the matrix $(\partial\zeta^\alpha/\partial z^\beta)$ defines A independently of the coordinate system.) The kernel of A contains E'. We assume that E' is exactly the kernel of A. Then denoting the image $AE\subset E$ by E'', we obtain a decomposition

$$E=E'\oplus E''.$$

Thus, E is *holomorphically* isomorphic to the direct sum of two holomorphic vector subbundles E' and E''. We denote the restriction of A to E'' by Λ. Then Λ is an automorphism of E''. Let h' and h'' be hermitian fibre metrics in E' and E'', respectively. We extend the hermitian fibre metric $h'+h''$ of E to a hermitian metric h of M.

We follow now closely § 6 of Chapter II. Let P be the bundle of unitary frames over M; it is a principal bundle over M with group $U(n)$.

Let P_N be the bundle of adapted frames over N; it is a principal bundle over N with group $U(n-r)\times U(r)$. (By an adapted frame we mean a unitary frame whose first $n-r$ basis elements are in E' and whose last r elements are in E''.) Considering the hermitian connection of M, let Ω be the curvature form on P. Then its restriction to P_N is of the form

$$\begin{pmatrix} \Omega_\tau & 0 \\ 0 & \Omega_\nu \end{pmatrix},$$

where, with an obvious identification, Ω_τ is the curvature form for h' and Ω_ν is the curvature form for h''.

Let ∇ denote the covariant differentiation of the hermitian connection defined by h. As in § 4, we write

$$\nabla Z = \nabla' Z + \nabla'' Z,$$

where $\nabla' Z$ and $\nabla'' Z$ are defined by the property that

$$\nabla'_{\bar{W}} Z = 0 \quad \text{and} \quad \nabla''_{W} Z = 0 \qquad \text{for all vectors } W \text{ of type } (1,0).$$

Then the endomorphism A of E coincides with the restriction of $\nabla' Z$ to N. As in § 6 of Chapter II, we may consider $\nabla' Z$ as a tensorial 0-form of type $\mathrm{ad}(U(n))$. Restricted to P_N, $\nabla' Z$ is of the form

$$\begin{pmatrix} 0 & 0 \\ 0 & \Lambda \end{pmatrix}.$$

Let f be an $\mathrm{ad}(U(n))$-invariant symmetric form of degree n on the Lie algebra $\mathfrak{u}(n)$. The simplest example is given by det (=determinant). As in § 6 of Chapter II, we define the residue $\mathrm{Res}_f(N)$ by

$$\mathrm{Res}_f(N)\cdot t^{n-r} = \int_N \frac{\bar{f}(t\Omega + \nabla' Z)}{\overline{\det}(t\Omega_\nu + \Lambda)},$$

where the bars over f and det indicate that the forms are pull down to N.

Now the theorem of Bott [1, 2] may be stated as follows:

Theorem 12.1. *Let M be a compact complex manifold of dimension n. Let Z be a holomorphic vector field of M with zero set* $\mathrm{Zero}(Z) = \bigcup N_i$, *where the N_i's are the connected components of* $\mathrm{Zero}(Z)$. *Let f be an* $(\mathrm{ad}\, U(n))$*-invariant symmetric form of degree n on $\mathfrak{u}(n)$. Then the characteristic number $\int_M \bar{f}(\Omega)$ of M defined by f is given by*

$$\int_M \bar{f}(\Omega) = \sum_i \mathrm{Res}_f(N_i),$$

provided that Z is non-degenerate along $\mathrm{Zero}(Z)$ *in the sense defined above.*

The proof is almost identical to that of Theorem 6.1 of Chapter II and hence is omitted. We remark that $A=\nabla' Z|_N$ is holomorphic and hence the coefficient of t^k in $\bar{f}(t\Omega+\nabla' Z)|_N$ is a polynomial in the curvature form with *constant* coefficients since a holomorphic function on a compact manifold must be constant. (In § 6 of Chapter II, a similar fact was derived from the property that ∇X is parallel on N.)

In Bott [1], the theorem is proved when Zero(Z) consists of isolated points. The general case is proved in Bott [2]. A generalization to meromorphic vector fields has been obtained by Baum-Bott [1]. For a different proof, see Atiyah-Singer [1]. When Zero(Z) consists of isolated points, the theorem follows also from a Lefschetz formula of Atiyah-Bott [1]. See also Illusie [1].

Corollary 12.2. *If a compact complex manifold admits a holomorphic vector field with empty zero set, then its Chern numbers vanish.*

IV. Affine, Conformal and Projective Transformations

1. The Group of Affine Transformations of an Affinely Connected Manifold

Let M be a manifold with an affine connection and $L(M)$ be the bundle of linear frames over M. Let θ and ω denote the canonical form and the connection form on $L(M)$, respectively. We recall (§ 1 of Chapter II) that a transformation f of M is said to be affine if the induced automorphism $\tilde{f}$ of $L(M)$ leaves ω invariant. We quote the following result established earlier (see Theorem 1.3 of Chapter II).

Theorem 1.1. *Let M be an n-dimensional manifold with an affine connection. Then the group $\mathfrak{A}(M)$ of affine transformations of M is a Lie transformation group of dimension $\leqq n(n+1)$.*

As in § 3 of Chapter II, it is natural to ask when the maximum dimension $n(n+1)$ is attained. We prove

Theorem 1.2. *Let M and $\mathfrak{A}(M)$ be as in Theorem* 1.1. *Then* $\dim \mathfrak{A}(M) = n(n+1)$ *if and only if M is an ordinary affine space with the natural flat affine connection.*

Proof. Assume $\dim \mathfrak{A}(M) = n(n+1)$. From Theorem 1.3 of Chapter II, it is clear that the identity component $\mathfrak{A}^0(M)$ acts simply transitively on each connected component of $L(M)$. This implies that every standard horizontal vector field $\tilde{X}$ (i.e., vector field $\tilde{X}$ such that $\omega(\tilde{X}) = 0$ and $\theta(\tilde{X}) = a$ constant element in $\mathbf{R}^n$) on $L(M)$ is complete; the proof is similar to that of Theorem 2.5 of Chapter II. This means that the connection is complete (see Proposition 6.5 of Chapter III in Kobayashi-Nomizu [1]).

Let $\mathfrak{A}_x$ be the isotropy subgroup of $\mathfrak{A}(M)$ at a point $x \in M$. Since $\dim \mathfrak{A}_x = n^2$, $\mathfrak{A}_x$ contains $\mathrm{GL}^+(n; \mathbf{R})$, where

$$\mathrm{GL}^+(n; \mathbf{R}) = \{a \in \mathrm{GL}(n; \mathbf{R});\ \det a > 0\}.$$

(Identifying $\mathfrak{A}_x$ with the linear isotropy group, we consider it as a subgroup of $\mathrm{GL}(n; \mathbf{R})$ by choosing a basis in the tangent space $T_x(M)$.) In particular, it contains homothetic transformations $t I_n$ with $t>0$. If K is a tensor of contravariant degree p and covariant degree q at the point x, the transformation $t I_n$ sends K into $t^{p-q} K$. Since the curvature tensor field R is of contravariant degree 1 and covariant degree 3, the transformation $t I_n$ sends R_x into $t^{-2} R_x$. On the other hand, $\mathfrak{A}_x$ must leave R_x invariant so that $R_x = t^{-2} R_x$ for all $t>0$. Hence, $R_x = 0$. This shows that the curvature tensor field vanishes identically. Similarly, the torsion tensor field vanishes identically. Hence, the connection is flat.

Let $\tilde{M}$ be the universal covering space of M. Since $\tilde{M}$ is a simply connected manifold with a complete flat affine connection, it is an ordinary affine space with the natural flat affine connection (see Theorem 7.8 of Chapter VI in Kobayashi-Nomizu [1]). From the fact that no element of $\mathfrak{A}(\tilde{M})$ other than the identity commutes with $\mathfrak{A}^0(\tilde{M})$ elementwise, we can conclude that M itself is simply connected as in the proof of Theorem 3.1 of Chapter II.

Finally, the converse is evident. q.e.d.

The local version of Theorem 1.2 is classical; see, for example, Eisenhart [1].

The following theorem is due to Egorov [1] (see also Yano [3]).

Theorem 1.3. *Let M be an n-dimensional manifold with an affine connection and $\mathfrak{A}(M)$ be the group of affine transformations. If* $\dim \mathfrak{A}(M) > n^2$, *then the connection has no torsion.*

Proof. Assuming that the torsion tensor does not vanish at a point x, we shall show that the dimension of the isotropy subgroup $\mathfrak{A}_x$ is at most $n^2 - n$. Since $\mathfrak{A}_x$ leaves the torsion tensor at x invariant, it suffices to prove the following algebraic lemma.

Lemma. *Let V be an n-dimensional vector space and T be a nonzero element of $V \otimes \Lambda^2 V^*$, where V^* is the dual space of V. Let G be the group of linear transformations of V leaving T invariant. Then*

$$\dim G \leqq n^2 - n.$$

Proof of Lemma. We consider T as a skew-symmetric bilinear mapping $V \times V \to V$. Let $X\colon V \to V$ be a linear transformation and $\varphi_t = e^{tX}$ be the 1-parameter group of linear transformations generated by X. Then X is in the Lie algebra $\mathfrak{g}$ of G if and only if

$$T(\varphi_t v, \varphi_t w) = \varphi_t(T(v, w)) \quad \text{for } v, w \in V \text{ and all } t \in \mathbf{R}.$$

Differentiating this equation with respect to t at $t=0$, we see that X is in $\mathfrak{g}$ if and only if

$$T(X v, w) + T(v, X w) = X(T(v, w)) \quad \text{for } v, w \in V.$$

Take a basis in V and let T^i_{jk} and X^i_j be the components of T and X with respect to the chosen basis. Then the equation above is equivalent to the following system of linear equations:

$$\sum_j (T^a_{jc} X^j_b + T^a_{bj} X^j_c - X^a_j T^j_{bc}) = 0 \qquad \text{for } a, b, c = 1, \ldots, n.$$

This can be rewritten as follows:

$$\sum_{i,j} A^{aj}_{bci} X^i_j = 0 \qquad (a, b, c = 1, \ldots, n),$$

where

$$A^{aj}_{bci} = T^a_{ic} \delta^j_b + T^a_{bi} \delta^j_c - T^j_{bc} \delta^a_i.$$

This is a system of linear equations with n^2 unknowns X^i_j and coefficients A^{aj}_{bci}. If we can find n linearly independent equations in this system, then we will know that the dimension of the space of solutions of this system does not exceed $n^2 - n$. Let T^a_{bc} be one of the non-vanishing components of T. Since T^i_{jk} is skew-symmetric in the lower indices, $T^a_{bc} \neq 0$ implies $b \neq c$. By reordering the basis if necessary, we may assume that either $a=1$, $b=2$, $c=3$, or $a=b=1$, $c=2$, i.e., $T^1_{23} \neq 0$ or $T^1_{12} \neq 0$.

We shall first consider the case where $T^1_{23} \neq 0$. We claim that the n linear equations given by

$$\sum_{i,j} A^{kj}_{23i} X^i_j = 0 \qquad k = 1, \ldots, n$$

are linearly independent. In fact, consider the $n \times n$ matrix (A^k_i) defined by $A^k_i = A^{k1}_{23i}$. Then

$$A^k_i = T^k_{i3} \delta^1_2 + T^k_{2i} \delta^1_3 - T^1_{23} \delta^k_i = -T^1_{23} \delta^k_i.$$

Since $T^1_{23} \neq 0$, the matrix (A^k_i) is non-singular.

Next, assume $T^1_{12} \neq 0$. We claim that the n linear equations given by

$$\sum_{i,j} A^{2j}_{12i} X^i_j = 0$$

$$\sum_{i,j} A^{1j}_{1ki} X^i_j = 0 \qquad k = 2, \ldots, n,$$

are linearly independent. In fact, consider the $n \times n$ matrix (A^j_k) defined by

$$A^j_1 = A^{2j}_{122} = T^2_{22} \delta^j_1 + T^2_{12} \delta^j_2 - T^j_{12} \delta^2_2 = T^2_{12} \delta^j_2 - T^j_{12},$$

$$A^j_k = A^{11}_{1k2} = T^1_{2k} \delta^j_1 + T^1_{12} \delta^j_k - T^j_{1k} \delta^1_2 = T^1_{2k} \delta^j_1 + T^1_{12} \delta^j_k$$

$$\text{for } j = 1, \ldots, n \quad \text{and} \quad k = 2, \ldots, n.$$

Then this matrix is of the following form:

$$\begin{pmatrix} -T^1_{12} & * \\ 0 & T^1_{12} I_{n-1} \end{pmatrix},$$

where I_{n-1} denotes the identity matrix of order $n-1$. This matrix is clearly non-singular. q.e.d.

We state another result of Egorov [3] which can be proved by a similar method (see also Yano [3]).

Theorem 1.4. *Let M and $\mathfrak{A}(M)$ be as in Theorem 1.3. If $\dim \mathfrak{A}(M) > n^2$, then the connection has neither torsion nor curvature, i.e., it is flat, provided $n \geqq 4$.*

Remark. In connection with Theorem 1.3, Egorov [5] (see also Yano [3]) has shown that if $\dim \mathfrak{A}(M) = n^2$ and $n \geqq 4$, then there exists a 1-form τ on M such that the torsion tensor T is given by

$$T(X, Y) = \tau(Y)X - \tau(X)Y \quad \text{for all vector fields } X, Y.$$

For example, define an affine connection in $M = \mathbf{R}^n$ in terms of the natural coordinate system $x^1, \ldots, x^n$ and the Christoffel symbols as follows:

$$\Gamma^i_{1k} = -\delta^i_k, \qquad \Gamma^i_{jk} = 0 \quad \text{for } j \neq 1,$$

so that its torsion T can be constructed from the 1-form $\tau = dx^1$ in the manner described above. Since the group $\mathfrak{A}(M)$ contains the translations in $\mathbf{R}^n$, it is transitive on $M = \mathbf{R}^n$ and its isotropy subgroup $\mathfrak{A}_0$ at the origin is the group of linear transformations of $\mathbf{R}^n$ leaving the point $x^1 = 1$, $x^2 = \cdots = x^n = 0$ invariant. Hence, $\dim \mathfrak{A}(M) = n^2$. Clearly, the curvature of this connection vanishes identically.

If we symmetrize the affine connection above by setting

$$\Gamma^k_{1k} = \Gamma^i_{k1} = -\tfrac{1}{2}\delta^i_k, \qquad \Gamma^i_{jk} = 0 \quad \text{for } j, k \neq 1,$$

then we obtain a torsionfree affine connection with nontrivial curvature on $M = \mathbf{R}^n$. The group $\mathfrak{A}(M)$ of affine transformations for this connection is the same group of dimension n^2.

Wang and Yano [1] have determined the n-dimensional affinely connected manifolds such that $\dim \mathfrak{A}(M) > n^2 - n + 5$. For more details on related results, see Yano [3].

2. Affine Transformations of Riemannian Manifolds

In this section we shall compare the group $\mathfrak{J}(M)$ of isometries and the group $\mathfrak{A}(M)$ of affine transformations of a Riemannian manifold M.

The following result is due to Hanno [1]. See also Kobayashi-Nomizu [1, vol. 1; p. 240].

Theorem 2.1. *Let $M = M_0 \times M_1 \times \cdots \times M_k$ be the de Rham decomposition of a complete, simply connected Riemannian manifold M, where M_0 is*

Euclidean and $M_1, \dots, M_k$ are all irreducible. Then

$$\mathfrak{A}^0(M) = \mathfrak{A}^0(M_0) \times \mathfrak{A}^0(M_1) \times \cdots \times \mathfrak{A}^0(M_k),$$
$$\mathfrak{J}^0(M) = \mathfrak{J}^0(M_0) \times \mathfrak{J}^0(M_1) \times \cdots \times \mathfrak{J}^0(M_k).$$

Since the group of affine transformations and that of isometries of a Eulcidean space are well understood, the study of $\mathfrak{A}^0(M)$ and $\mathfrak{J}^0(M)$ is essentially reduced to the case where M is irreducible. For a proof of the following result, the reader is again referred to Kobayashi-Nomizu [1, vol. 1; p. 242].

Theorem 2.2. *If M is a complete, irreducible Riemannian manifold, then $\mathfrak{A}(M) = \mathfrak{J}(M)$ except when M is a 1-dimensional Euclidean space.*

Let M be a complete Riemannian manifold and $\tilde{M}$ be its universal covering space. Let $\tilde{M} = M_0 \times M_1 \times \cdots \times M_k$ be the de Rham decomposition of $\tilde{M}$. By Theorem 2.1, the Lie algebra $\mathfrak{a}(\tilde{M})$ of $\mathfrak{A}(\tilde{M})$ is isomorphic to $\mathfrak{a}(M_0) + \mathfrak{a}(M_1) + \cdots + \mathfrak{a}(M_k)$. Let X be an infinitesimal affine transformation of M and $\tilde{X}$ be its natural lift to $\tilde{M}$. Let $(X_0, X_1, \dots, X_k)$ be the element of $\mathfrak{a}(M_0) + \mathfrak{a}(M_1) + \cdots + \mathfrak{a}(M_k)$ corresponding to $\tilde{X} \in \mathfrak{a}(\tilde{M})$. By Theorem 2.2, $X_1, \dots, X_k$ are all infinitesimal isometries. If $\tilde{M}$ has no Euclidean factor M_0, then X_0 is zero so that $\tilde{X}$ is an infinitesimal isometry. If $\tilde{X}$ is an infinitesimal isometry, so is X. The assumption that M_0 be trivial can be expressed in terms of the restricted linear holonomy group. We can therefore state the result as follows (see also Lichnerowicz [1; p. 83], Yano-Nagano [3]).

Corollary 2.3. *If M is a complete Riemannian manifold such that its restricted linear holonomy group $\Psi^0(x)$ leaves no nonzero vector at x fixed, then $\mathfrak{A}^0(M) = \mathfrak{J}^0(M)$.*

The assumption in Corollary 2.3 is technically a little stronger than the condition that M admits no nonzero parallel vector field. The latter condition amounts to assuming that the linear holonomy group $\Psi(x)$ leaves no nonzero vector at x fixed.

With the notation above, if the length of an infinitesimal affine transformation X of M is bounded, the same is true for $X_0, X_1, \dots, X_k$. But an infinitesimal affine transformation X_0 of the Euclidean space M_0 is of bounded length if and only if X_0 is an infinitesimal translation which is a very special kind of infinitesimal isometry (see Kobayashi-Nomizu [1, vol. 1; p. 244]). Hence (Hano [1]),

Corollary 2.4. *If X is an infinitesimal affine transformation of a complete Riemannian manifold with bounded length, then it is an infinitesimal isometry.*

In particular, we rediscover Corollary 2.4 of Chapter II due to Yano.

Corollary 2.5. *If M is a compact Riemannian manifold, then* $\mathfrak{A}^0(M) = \mathfrak{I}^0(M)$.

3. Cartan Connections

In this section we shall treat conformal and projective connections in a unified manner.

Let M be a manifold of dimension n, $\mathfrak{L}$ a Lie group, $\mathfrak{L}_0$ a closed subgroup of $\mathfrak{L}$ with $\dim \mathfrak{L}/\mathfrak{L}_0 = n$ and P a principal bundle over M with group $\mathfrak{L}_0$. We give a few examples:

Example 3.1. Let $\mathfrak{L}$ be a Lie group and $\mathfrak{L}_0$ a closed subgroup of $\mathfrak{L}$. Set $M = \mathfrak{L}/\mathfrak{L}_0$ and $P = \mathfrak{L}$.

Example 3.2. Let $\mathfrak{L}$ be the affine group $\mathfrak{A}(n)$ acting on an n-dimensional affine space and $\mathfrak{L}_0 = GL(n; \mathbf{R})$ an isotropy subgroup of $\mathfrak{L}$ so that $\mathfrak{L}/\mathfrak{L}_0$ is the affine space. Let M be a manifold of dimension n and P the bundle of linear frames over M.

Example 3.3. Let $\mathfrak{L}$ be the group of Euclidean motions of an n-dimensional Euclidean space and $\mathfrak{L}_0 = O(n)$ an isotropy subgroup of $\mathfrak{L}$ so that $\mathfrak{L}/\mathfrak{L}_0$ is the Euclidean space. Let M be a Riemannian manifold of dimension n and P the bundle of orthonormal frames over M.

Example 3.4. Let $\mathfrak{L}$ be the projective general linear group $PGL(n; \mathbf{R})$ acting on an n-dimensional real projective space and $\mathfrak{L}_0$ an isotropy subgroup of $\mathfrak{L}$ so that $\mathfrak{L}/\mathfrak{L}_0$ is the projective space. Let M be an n-dimensional manifold. We shall later construct a principal bundle P over M with group $\mathfrak{L}_0$, called a projective structure.

Example 3.5. Let $\mathfrak{L}$ be the Möbius group $O(n+1, 1)$ acting on an n-dimensional sphere S^n and $\mathfrak{L}_0$ an isotropy subgroup of $\mathfrak{L}$ so that $\mathfrak{L}/\mathfrak{L}_0$ is the sphere S^n. (This will be explained in detail later.) Let M be an n-dimensional manifold with a conformal structure and P be the first prolongation of the conformal structure. (This will be also explained in detail later.)

Since $\mathfrak{L}_0$ acts on P on the right, every element A of the Lie algebra $\mathfrak{l}_0$ of $\mathfrak{L}_0$ defines a vertical vector field on P, called the *fundamental vector field* corresponding to A (Kobayashi-Nomizu [1, vol. 1; p. 51]). This vector field will be denoted by A^*. For each element a of $\mathfrak{L}_0$, the right translation by a acting on P will be denoted by R_a. A *Cartan connection* in the bundle P is a 1-form ω on P with values in the Lie algebra $\mathfrak{l}$ of $\mathfrak{L}$ satisfying the following conditions:

(a) $\omega(A^*)=A$ for every $A \in \mathfrak{l}_0$;

(b) $(R_a)^* \omega = \mathrm{ad}(a^{-1})\omega$ for every element $a \in \mathfrak{L}_0$, where $\mathrm{ad}(a^{-1})$ is the adjoint action of a^{-1} on $\mathfrak{l}$;

(c) $\omega(X) \neq 0$ for every nonzero vector X of P.

Condition (c) means that ω defines a linear isomorphism of the tangent space $T_u(P)$ onto the Lie algebra $\mathfrak{l}$ for every $u \in P$ since $\dim P = \dim \mathfrak{L}$. In other words, ω defines an absolute parallelism on P.

A Cartan connection in P is not a connection in P in the usual sense, for ω is not $\mathfrak{l}_0$-valued. It can be, however, considered as a connection in a larger bundle $P^{\mathfrak{L}}$ obtained by enlarging the structure group of P to $\mathfrak{L}$, i.e.,

$$P^{\mathfrak{L}} = P \times_{\mathfrak{L}_0} \mathfrak{L}.$$

Then P is a subbundle of $P^{\mathfrak{L}}$ and a Cartan connection ω in P can be uniquely extended to a usual connection form on $P^{\mathfrak{L}}$, also denoted by ω. (Take, for instance, Example 3.2. Then an affine connection of M is a Cartan connection in P. On the other hand, a linear connection of M is an ordinary connection in P. For more details on this point, see Kobayashi-Nomizu [1; Chapter III, §§ 2–3].)

If we set $E = P^{\mathfrak{L}}/\mathfrak{L}_0$, then E is the bundle with fibre $\mathfrak{L}/\mathfrak{L}_0$ associated with P. We can identify M with the image of the natural mapping $P \to P^{\mathfrak{L}}/\mathfrak{L}_0$. In other words, we have a natural cross section $M \to E$. In Examples 3.2 and 3.3, E is the tangent bundle of M and the natural cross section is the zero section. In Example 3.1, E is the product bundle $M \times M$ and the natural cross section $M \to E = M \times M$ is the diagonal map. Geometrically speaking, condition (c) means that the fibre of E over each point $x \in M$ is tangent to M at x, see Ehresmann [1], Kobayashi [4]. But this geometric interpretation of (c) will not be used here.

We define the curvature form Ω of the Cartan connection ω by the following structure equation:

$$d\omega = -\tfrac{1}{2}[\omega, \omega] + \Omega.$$

It is an $\mathfrak{l}$-valued 2-form on P. For instance, if ω is an affine connection, then Ω is a 2-form with values in the Lie algebra $\mathfrak{a}(n)$ of the group $\mathfrak{A}(n)$ of affine motions. Decompose $\mathfrak{a}(n)$ into the vector space direct sum of the translation part $\mathbf{R}^n$ and the linear transformation part $\mathfrak{gl}(n; \mathbf{R})$. Then the $\mathbf{R}^n$-component of Ω is the usual torsion form and the $\mathfrak{gl}(n; \mathbf{R})$-component of Ω is the usual curvature form of the corresponding linear connection, see Kobayashi-Nomizu [1; § 3 of Chapter III].

Given a Cartan connection ω in P, we call a transformation φ of P an *automorphism* of (P, ω) if it is a bundle automorphism, i.e., commutes with the right translations R_a, $(a \in \mathfrak{L}_0)$, and if it preserves the form ω.

From Theorem 3.2 of Chapter I we obtain the following result (Kobayashi [4]):

Theorem 3.1. *Let ω be a Cartan connection in P. Then the group $\mathfrak{A}(P,\omega)$ of automorphisms of (P,ω) is a Lie group with* $\dim \mathfrak{A}(P,\omega) \leqq \dim P$.

For a fixed point u_0 of P, the mapping $\varphi \in \mathfrak{A}(P,\omega) \to \varphi(u_0) \in P$ imbeds $\mathfrak{A}(P,\omega)$ as a closed submanifold of P.

From now on we shall assume that the Lie algebra $\mathfrak{l}$ of $\mathfrak{L}$ is graded as follows:

$$\mathfrak{l} = \mathfrak{g}_{-1} + \mathfrak{g}_0 + \mathfrak{g}_1 + \cdots + \mathfrak{g}_k \quad \text{(a vector space direct sum)},$$

with

$$[\mathfrak{g}_i, \mathfrak{g}_j] \subset \mathfrak{g}_{i+j} \quad \text{and} \quad \mathfrak{l}_0 = \mathfrak{g}_0 + \mathfrak{g}_1 + \cdots + \mathfrak{g}_k.$$

Let

$$\omega = \omega_{-1} + \omega_0 + \omega_1 + \cdots + \omega_k$$

be the corresponding decomposition of the Cartan connection ω. Since ω defines an absolute parallelism on P, the algebra of differential forms on P is generated by ω (i.e., the components of ω with respect to a basis for $\mathfrak{l}$) and functions on P. But *the curvature Ω of ω does not involve ω_0, $\omega_1, \ldots, \omega_k$*. In fact, this is a direct consequence of the following three facts:

(i) The $\mathfrak{g}_{-1}$-component ω_{-1}, restricted to each fibre of P, vanishes identically.

(ii) The $\mathfrak{l}_0$-component $\omega_0 + \omega_1 + \cdots + \omega_k$, restricted to each fibre of P, is the Maurer-Cartan form and hence defines an absolute parallelism.

(iii) The curvature Ω, restricted to each fibre of P, vanishes identically.

Condition (a) for Cartan connections implies both (i) and (ii). To prove (iii), it suffices to observe that the structure equation of the connection restricted to a fibre gives the structure equation of Maurer-Cartan for the group $\mathfrak{L}_0$ and then apply (i).

If we choose a basis $e_1, \ldots, e_n$ for $\mathfrak{g}_{-1}$ and write

$$\omega_{-1} = \omega^1 e_1 + \cdots + \omega^n e_n,$$

then we can express the curvature Ω of the Cartan connection ω as follows:

$$\Omega = \sum \tfrac{1}{2} K_{ij}\, \omega^i \wedge \omega^j,$$

where each K_{ij} is an $\mathfrak{l}$-valued function on P.

Theorem 3.2. *Let $\mathfrak{L}$ be a Lie group and $\mathfrak{L}_0$ a closed subgroup of $\mathfrak{L}$ such that the Lie algebra $\mathfrak{l}_0$ of $\mathfrak{L}_0$ contains an element E with the property that the Lie algebra $\mathfrak{l}$ of $\mathfrak{L}$ is graded as follows:*

$$\mathfrak{l} = \mathfrak{g}_{-1} + \mathfrak{g}_0 + \mathfrak{g}_1 + \cdots + \mathfrak{g}_k,$$

where

$$\mathfrak{g}_r = \{X \in \mathfrak{l};\ [E, X] = rX\} \quad \textit{for } r = -1, 0, 1, \ldots, k.$$

Let P be a principal $\mathfrak{L}_0$-bundle over M and ω a Cartan connection in P (with values in $\mathfrak{l}$). If the automorphism group $\mathfrak{A}(P, \omega)$ has the maximum dimension, i.e., $\dim \mathfrak{A}(P, \omega) = \dim P$, *then the curvature Ω of ω vanishes identically.*

Proof. Set $a_t = \exp(tE) \in \mathfrak{L}_0$. Fix a point $u_0 \in P$ and let φ_t be an element of $\mathfrak{A}(P, \omega)$ such that $R_{a_t} u_0 = \varphi_t(u_0)$. Since the orbit of $\mathfrak{A}(P, \omega)$ through u_0 is closed and has the same dimension as P, such an element φ_t exists. Let Ω_r be the $\mathfrak{g}_r$-component of Ω and write

$$\Omega_r = \sum \tfrac{1}{2} K_{rij}\, \omega^i \wedge \omega^j.$$

Since φ_t leaves ω^i and Ω_r invariant, it leaves K_{rij} invariant. We shall compare this action of φ_t with that of R_{a_t}. Since $\mathrm{ad}(E)$ coincides with the multiplication by r on $\mathfrak{g}_r$, $\mathrm{ad}(a_t)$ is the multiplication by e^{rt} on $\mathfrak{g}_r$. Hence,

$$R^*_{a_t}(\Omega_r) = \mathrm{ad}(a_t^{-1})\, \Omega_r = e^{-rt}\, \Omega_r,$$
$$R^*_{a_t}(\omega_{-1}) = \mathrm{ad}(a_t^{-1})\, \omega_{-1} = e^t\, \omega_{-1}.$$

From these two equalities, we obtain

$$\begin{aligned} \sum \tfrac{1}{2} e^{-rt} K_{rij}\, \omega^i \wedge \omega^j &= e^{-rt}\, \Omega_r = R^*_{a_t}(\Omega_r) \\ &= \sum \tfrac{1}{2} R^*_{a_t}(K_{rij})\, R^*_{a_t}\omega^i \wedge R^*_{a_t}\omega^j \\ &= \sum \tfrac{1}{2} R^*_{a_t}(K_{rij})\, e^{2t}\, \omega^i \wedge \omega^j. \end{aligned}$$

Hence,

$$R^*_{a_t}(K_{rij}) = e^{-(r+2)t} K_{rij}.$$

On the other hand,

$$\varphi_t^*(K_{rij}) = K_{rij}.$$

Comparing these two equalities at u_0, we obtain

$$K_{rij}(u_0) = K_{rij}(\varphi_t u_0) = K_{rij}(R_{a_t} u_0) = e^{-(r+2)t} K_{rij}(u_0).$$

Hence, $K_{rij}(u_0) = 0$. Since u_0 is an arbitrary point of P, this proves our assertion. q.e.d.

If we apply Theorem 3.2 to Examples 3.2 and 3.3, we obtain local versions of Theorem 3.1 of Chapter II and Theorem 1.2. (Note that $\mathfrak{l} = \mathfrak{g}_{-1} + \mathfrak{g}_0$ in Examples 3.2 and 3.3 and the $\mathfrak{g}_{-1}$-component of Ω is the torsion form while the $\mathfrak{g}_0$-component of Ω is the curvature form in the usual sense.)

For this section, see also Ogiue [2].

4. Projective and Conformal Connections

In the preceding section, we studied Cartan connections in general. In this section we shall describe both projective and conformal connections in a unified manner. We begin with a simple algebraic proposition (see Kobayashi-Nagano [1]).

Proposition 4.1. *Let* $\mathfrak{l}=\mathfrak{g}_{-1}+\mathfrak{g}_0+\mathfrak{g}_1+\cdots+\mathfrak{g}_k$ *be a semisimple graded Lie algebra (with* $\mathfrak{g}_{-1}\neq 0$ *and* $\mathfrak{g}_0\neq 0$*). Then*

(1) $\mathfrak{l}=\mathfrak{g}_{-1}+\mathfrak{g}_0+\mathfrak{g}_1$, *i.e.,* $\mathfrak{g}_2=0$.

(2) *The linear endomorphism* α *of* $\mathfrak{l}$ *defined by*

$$\alpha(X_{-1}+X_0+X_1)=-X_{-1}+X_0-X_1 \qquad \text{for } X_i\in\mathfrak{g}_i$$

is an involutive automorphism of $\mathfrak{l}$.

(3) *With respect to the Killing-Cartan form* B *of* $\mathfrak{l}$,

(i) $\mathfrak{g}_{-1}+\mathfrak{g}_1$ *is perpendicular to* $\mathfrak{g}_0$;

(ii) $B|\mathfrak{g}_{-1}=0$ *and* $B|\mathfrak{g}_1=0$;

(iii) $\mathfrak{g}_1$ *is the dual vector space of* $\mathfrak{g}_{-1}$ *under the dual pairing* $(X_{-1}, X_1)\rightarrow B(X_{-1}, X_1)$.

(4) *The two representations* $\mathrm{ad}_{\mathfrak{l}}(\mathfrak{g}_0)|\mathfrak{g}_{-1}$ *and* $\mathrm{ad}_{\mathfrak{l}}(\mathfrak{g}_0)|\mathfrak{g}_1$ *of* $\mathfrak{g}_0$ *are dual to each other with respect to the Killing-Cartan form* B.

(5) *There is a unique element* $E\in\mathfrak{g}_0$ *such that*

$$\mathfrak{g}_i=\{X\in\mathfrak{l};\ [E,X]=iX\} \qquad i=-1,0,1.$$

Proof. (1) Let $X\in\mathfrak{g}_i$ and $Z\in\mathfrak{g}_2$. Since $(\mathrm{ad}\,X)(\mathrm{ad}\,Z)$ maps $\mathfrak{g}_j$ into $\mathfrak{g}_{j+i+2}$ and $i+2\geqq 1$ so that $j\neq j+i+2$, the trace of $(\mathrm{ad}\,X)$ $(\mathrm{ad}\,Z)$ is zero. Hence, $B(X,Z)=0$. Since B is non-degenerate, it follows that $Z=0$.

(2) The proof is straightforward.

(3) Since B is invariant by the automorphism α, we have

$$B(X_{-1}+X_1, X_0)=B\big(\alpha(X_{-1}+X_1), \alpha(X_0)\big)=-B(X_{-1}+X_1, X_0),$$

which proves (i). To prove (ii), let $X_{-1}, Y_{-1}\in\mathfrak{g}_{-1}$. Then $(\mathrm{ad}\,X_{-1})(\mathrm{ad}\,Y_{-1})$ maps $\mathfrak{g}_i$ into $\mathfrak{g}_{i-2}$ and hence its trace is zero. This shows that

$$B(X_{-1}, Y_{-1})=0.$$

Similarly, $B|\mathfrak{g}_1=0$. To prove (iii), let $X_{-1}\in\mathfrak{g}_{-1}$ and assume $B(X_{-1}, \mathfrak{g}_1)=0$. By (i) and (ii), this implies $B(X_{-1}, \mathfrak{l})=0$. Hence, $X_{-1}=0$. Similarly, if $X_1\in\mathfrak{g}_1$ and $B(X_1, \mathfrak{g}_{-1})=0$, then $X_1=0$.

(4) Let $X_0\in\mathfrak{g}_0$. Since B is invariant by $\mathrm{ad}\,X_0$, we have

$$B\big(\mathrm{ad}(X_0)X_{-1}, X_1\big)=-B\big(X_{-1}, \mathrm{ad}(X_0)X_1\big) \qquad \text{for } X_i\in\mathfrak{g}_i,\ i=\pm 1.$$

(5) Let ε be the linear endomorphism of $\mathfrak{l}$ defined by

$$\varepsilon(X_i) = iX_i \quad \text{for } X_i \in \mathfrak{g}_i, \quad i = -1, 0, 1.$$

Then ε is a derivation of the Lie algebra. Since every derivation of a semi-simple Lie algebra is inner, there is a unique element $E \in \mathfrak{l}$ such that $\mathrm{ad}(E) = \varepsilon$. To prove that E is in $\mathfrak{g}_0$, we observe that $\mathrm{ad}(\alpha(E))$ coincides with $\mathrm{ad}(E)$. Hence, $\alpha(E) = E$. By the definition of α, $E \in \mathfrak{g}_0$. q.e.d.

The graded Lie algebras of Proposition 4.1 have been classified by Kobayashi-Nagano [1]. We are interested here only in the following two example. For a general theory, see Ochiai [3].

Example 4.1. Let $\mathfrak{L}/\mathfrak{L}_0$ be the real projective space of dimension n, where

$$\mathfrak{L} = \mathrm{PGL}(n; \mathbf{R}) = \mathrm{SL}(n+1; \mathbf{R})/\text{center},$$

$$\mathfrak{L}_0 = \left\{ \begin{pmatrix} A & 0 \\ \xi & a \end{pmatrix} \in \mathrm{SL}(n+1; \mathbf{R}) \right\} / \text{center},$$

where $A \in \mathrm{GL}(n; \mathbf{R})$ and ξ is a row n-vector,

$$\mathfrak{L}_1 = \left\{ \begin{pmatrix} I_n & 0 \\ \xi & 1 \end{pmatrix}; \ \xi\text{: row } n\text{-vector} \right\}.$$

The graded Lie algebra $\mathfrak{l} = \mathfrak{g}_{-1} + \mathfrak{g}_0 + \mathfrak{g}_1$ with this $\mathfrak{L}/\mathfrak{L}_0$ is given by

$$\mathfrak{l} = \mathfrak{sl}(n+1; \mathbf{R}),$$

$$\mathfrak{g}_{-1} = \left\{ \begin{pmatrix} 0 & v \\ 0 & 0 \end{pmatrix} \right\}, \quad \mathfrak{g}_0 = \left\{ \begin{pmatrix} A & 0 \\ 0 & a \end{pmatrix}; \ \text{trace } A + a = 0 \right\}, \quad \mathfrak{g}_1 = \left\{ \begin{pmatrix} 0 & 0 \\ \xi & 0 \end{pmatrix} \right\},$$

where v is a column n-vector, ξ is a row n-vector, $A \in \mathfrak{gl}(n; \mathbf{R})$ and $a \in \mathbf{R}$. The element $E \in \mathfrak{g}_0$ is given by

$$E = \begin{pmatrix} aI_n & 0 \\ 0 & b \end{pmatrix}, \quad \text{where } a = -1/n+1, \ b = n/n+1.$$

We may describe the graded Lie algebra $\mathfrak{l}$ as follows. Let V be the n-dimensional vector space of column n-vectors and V^* be the dual space consisting of row n-vectors. Then

$$\mathfrak{l} = V + \mathfrak{gl}(n; \mathbf{R}) + V^*$$

under the identification

$$\begin{pmatrix} 0 & v \\ 0 & 0 \end{pmatrix} \in \mathfrak{g}_{-1} \to v \in V^*, \quad \begin{pmatrix} 0 & 0 \\ \xi & 0 \end{pmatrix} \in \mathfrak{g}_1 \to \xi \in V^*,$$

$$\begin{pmatrix} A & 0 \\ 0 & a \end{pmatrix} \in \mathfrak{g}_0 \to A - aI_n \in \mathfrak{gl}(n; \mathbf{R}).$$

Under this identification, the Lie algebra structure in $\mathfrak{l}$ is given by

$$[v, v']=0, \quad [\xi, \xi']=0, \quad [U, v]=Uv, \quad [\xi, U]=\xi U,$$
$$[U, U']=UU'-U'U, \quad [v, \xi]=v\xi+\xi v I_n,$$

where $v, v' \in V$, $\xi, \xi' \in V^*$, $U, U' \in \mathfrak{gl}(n; \mathbf{R})$. Clearly, the element E is now given by $-I_n \in \mathfrak{gl}(n; \mathbf{R})$. With respect to the natural bases $e_1, \ldots, e_n$ of V, $e^1, \ldots, e^n$ of V^* and e_i^j of $\mathfrak{gl}(n; \mathbf{R})$, we write the Maurer-Cartan form ω of $\mathrm{PGL}(n; \mathbf{R})$ as follows:

$$\omega = \sum \omega^i e_i + \sum \omega^i_j e^j_i + \sum \omega_j e^j.$$

Then the Maurer-Cartan structure equations of $\mathrm{PGL}(n; \mathbf{R})$ are given by

$$d\omega^i = -\sum \omega^i_k \wedge \omega^k,$$
$$d\omega^i_j = -\sum \omega^i_k \wedge \omega^k_j - \omega^i \wedge \omega_j + \delta^i_j \sum \omega_k \wedge \omega^k,$$
$$d\omega_j = -\sum \omega_k \wedge \omega^k_j.$$

Example 4.2. We describe the n-dimensional Möbius space S^n, first geometrically and then group-theoretically. Let S be the symmetric matrix of order $n+2$ given by

$$S = \begin{pmatrix} 0 & 0 & -1 \\ 0 & I_n & 0 \\ -1 & 0 & 0 \end{pmatrix}.$$

Let $x \in \mathbf{R}^{n+2}$ be a nonzero column vector, considered as a point in the real projective space $P_{n+1}(\mathbf{R})$. The quadric S^n in $P_{n+1}(\mathbf{R})$ defined by

$${}^t x S x = 0$$

is the n-dimensional *Möbius space.* It is diffeomorphic to the n-sphere $(y^1)^2 + \cdots + (y^{n+1})^2 = 1$ in $\mathbf{R}^{n+1}$ under the mapping defined in terms of the natural homogeneous coordinate system $x^0, x^1, \ldots, x^{n+1}$ of $P_{n+1}(\mathbf{R})$ by

$$x^0 = \tfrac{1}{2}(1-y^{n+1}), \quad x^1 = y^1, \ldots, x^n = y^n, \quad x^{n+1} = \tfrac{1}{2}(1+y^{n+1}).$$

Let p be the natural projection from $\mathbf{R}^{n+2} - \{0\}$ onto $P_{n+1}(\mathbf{R})$ and ds^2 be the natural Riemannian metric on $P_{n+1}(\mathbf{R})$ defined by

$$p^*(ds^2) = 2\{(\sum x^i x^i)(\sum dx^i dx^i) - (\sum x^i dx^i)^2\}/(\sum x^i x^i)^2.$$

Then the diffeomorphism above is an isometric mapping of the unit sphere onto the Möbius space S^n. Let

$$\mathfrak{L} = \mathrm{O}(n+1, 1) = \{X \in \mathrm{GL}(n+2; \mathbf{R});\ {}^t X S X = S\},$$

$$\mathfrak{L}_0 = \left\{ \begin{pmatrix} a^{-1} & 0 & 0 \\ v & A & 0 \\ b & \xi & a \end{pmatrix} \in \mathrm{O}(n+1, 1);\ A \in \mathrm{O}(n)\ a \in \mathbf{R},\ \xi \in \mathbf{R}^n \right\},$$

where ξ is written as a row vector. By a simple calculation it can be shown that

$$v = a^{-1} A \cdot {}^t\xi, \qquad b = (2a)^{-1} \xi \cdot {}^t\xi.$$

Let

$$\mathfrak{L}_1 = \left\{ \begin{pmatrix} 1 & 0 & 0 \\ {}^t\xi & I_n & 0 \\ b & \xi & 1 \end{pmatrix} \in O(n+1, 1);\ \xi \in \mathbf{R}^n \right\}, \qquad b = \tfrac{1}{2} \xi \cdot {}^t\xi.$$

Then $\mathfrak{L}$ acts transitively on the Möbius space S^n with $\mathfrak{L}_0$ as the isotropy subgroup at the point defined by $x^0 = x^1 = \cdots = x^n = 0$, called the origin of the Möbius space S^n. The subgroup $\mathfrak{L}_1$ is the kernel of the linear isotropy representation of $\mathfrak{L}_0$ at the origin. The graded Lie algebra $\mathfrak{l} = \mathfrak{g}_{-1} + \mathfrak{g}_0 + \mathfrak{g}_1$ associated with $\mathfrak{L}/\mathfrak{L}_0$ is given by

$$\mathfrak{l} = \mathfrak{o}(n+1, 1) = \{ X \in \mathfrak{gl}(n+2; \mathbf{R});\ {}^tXS + SX = 0 \},$$

$$\mathfrak{g}_{-1} = \left\{ \begin{pmatrix} 0 & {}^tv & 0 \\ 0 & 0 & v \\ 0 & 0 & 0 \end{pmatrix} \right\}, \qquad \mathfrak{g}_0 = \left\{ \begin{pmatrix} -a & 0 & 0 \\ 0 & A & 0 \\ 0 & 0 & a \end{pmatrix};\ A \in \mathfrak{o}(n) \right\},$$

$$\mathfrak{g}_1 = \left\{ \begin{pmatrix} 0 & 0 & 0 \\ {}^t\xi & 0 & 0 \\ 0 & \xi & 0 \end{pmatrix} \right\},$$

where v is a column n-vector, ξ is a row n-vector and $a \in \mathbf{R}$. The element $E \in \mathfrak{g}_0$ is given by

$$E = \begin{pmatrix} -1 & 0 & 0 \\ 0 & 0 & 0 \\ 0 & 0 & 1 \end{pmatrix}.$$

As in Example 4.1, we can describe the graded Lie algebra $\mathfrak{l}$ as

$$\mathfrak{l} = V + \mathfrak{co}(n) + V^*$$

under the identification

$$\begin{pmatrix} 0 & {}^tv & 0 \\ 0 & 0 & v \\ 0 & 0 & 0 \end{pmatrix} \in \mathfrak{g}_{-1} \to v \in V, \qquad \begin{pmatrix} 0 & 0 & 0 \\ {}^t\xi & 0 & 0 \\ 0 & \xi & 0 \end{pmatrix} \in \mathfrak{g}_1 \to \xi \in V^*,$$

$$\begin{pmatrix} -a & 0 & 0 \\ 0 & A & 0 \\ 0 & 0 & a \end{pmatrix} \in \mathfrak{g}_0 \to A - a I_n \in \mathfrak{co}(n).$$

Under this identification, the Lie algebra structure in $\mathfrak{l}$ is given by

$$[v, v] = 0, \quad [\xi, \xi'] = 0, \quad [U, v] = Uv, \quad [\xi, U] = \xi U,$$
$$[U, U'] = UU' - U'U, \quad [v, \xi] = v\xi - {}^t(v\xi) + (\xi v) I_n,$$

where $v, v' \in V$, $\xi, \xi' \in V^*$, $U, U' \in \mathfrak{co}(n)$. Clearly, the element E is now given by $-I_n \in \mathfrak{co}(n)$. Using the same bases for V, V^* and $\mathfrak{gl}(n; \mathbf{R})$ as in Example 4.1, we can write the Maurer-Cartan form ω of $O(n+1, 1)$ as follows:

$$\omega = \sum \omega^i e_i + \sum \omega^i_j e^j_i + \sum \omega_j e^j,$$

where (ω^i_j) is $\mathfrak{co}(n)$-valued. Then the Maurer-Cartan structure equations of $O(n+1, 1)$ are given by

$$d\omega^i = -\sum \omega^i_k \wedge \omega^k,$$
$$d\omega^i_j = -\sum \omega^i_k \wedge \omega^k_j - \omega^i \wedge \omega_j - \omega_i \wedge \omega^j + \delta^i_j \sum \omega_k \wedge \omega^k,$$
$$d\omega_j = -\sum \omega_k \wedge \omega^k_j.$$

Let P be a principal $\mathfrak{L}_0$-bundle over M. Given a $\mathfrak{g}_{-1}$-valued 1-form ω_{-1} and $\mathfrak{g}_0$-valued 1-form ω_0, we consider the question whether there exists a natural $\mathfrak{g}_1$-valued 1-form ω_1 such that $\omega = \omega_{-1} + \omega_0 + \omega_1$ is a Cartan connection in P. There are some obvious conditions which must be imposed on ω_{-1} and ω_0. Corresponding to conditions (a), (b) and (c) for Cartan connections stated in § 3, we must have the following:

(a′) $\quad \omega_{-1}(A^*) = 0$ and $\omega_0(A^*) = A_0$ for every $A \in \mathfrak{g}_0 + \mathfrak{g}_1$,

where A_0 is the $\mathfrak{g}_0$-component of A;

(b′) $\quad (R_a)^*(\omega_{-1} + \omega_0) = (\mathrm{ad}\, a^{-1})(\omega_{-1} + \omega_0)$ for every $a \in \mathfrak{L}_0$,

where $\mathrm{ad}\, a^{-1}$ is the transformation of $\mathfrak{g}_{-1} + \mathfrak{g}_0$ $(= \mathfrak{l}/\mathfrak{g}_1)$ induced by $\mathrm{ad}\, a^{-1}$: $\mathfrak{l} \to \mathfrak{l}$;

(c′) a tangent vector X of P is vertical (i..e., tangent to a fibre) if $\omega_{-1}(X) = 0$.

We are now in a position to state a theorem on normal projective and conformal connections.

Theorem 4.2. *Let $\mathfrak{L}/\mathfrak{L}_0$ be as in either Example* 4.1 *or Example* 4.2 *and P be a principal $\mathfrak{L}_0$-bundle over a manifold M of dimension n ($\geqq 3$ for Ex.* 4.2*). Given a $\mathfrak{g}_{-1}$-valued 1-form $\omega_{-1} = (\omega^i)$ and a $\mathfrak{g}_0$-valued 1-form $\omega_0 = (\omega^i_j)$ on P satisfying conditions (a′), (b′), (c′) and*

(I) $$d\omega^i = -\sum \omega^i_k \wedge \omega^k,$$

there is a unique Cartan connection $\omega=\omega_{-1}+\omega_0+\omega_1=(\omega^i;\omega^i_j;\omega_j)$ *such that the curvature* $\Omega=(0;\Omega^i_j;\Omega_j)$ *satisfies the following condition:*

$$\sum K^i_{jil}=0, \qquad \textit{where } \Omega^i_j=\sum \tfrac{1}{2} K^i_{jkl}\,\omega^k\wedge\omega^l.$$

This unique connection is called a *normal projective* or *conformal connection* according as $\mathfrak{L}/\mathfrak{L}_0$ is as in Example 4.1 or 4.2.

Proof. Let $\omega=(\omega^i;\omega^i_j;\omega_j)$ be a Cartan connection with the given $(\omega^i;\omega^i_j)$. In addition to the first structure equation (I) above, we have

$$(\mathrm{II})_p \qquad d\omega^i_j=-\sum\omega^i_k\wedge\omega^k_j-\omega^i\wedge\omega_j+\delta^i_j\sum\omega_k\wedge\omega^k+\Omega^i_j,$$

or

$$(\mathrm{II})_c \qquad d\omega^i_j=-\sum\omega^i_k\wedge\omega^k_j-\omega^i\wedge\omega_j-\omega_i\wedge\omega^j+\delta^i_j\sum\omega_k\wedge\omega^k+\Omega^i_j,$$

according as the connection is projective or conformal, and also

$$(\mathrm{III}) \qquad d\omega_j=-\sum\omega_k\wedge\omega^k_j+\Omega_j.$$

Applying exterior differentiation d to (I), making use of (I) and (II) and collecting the terms not involving ω^i_j and ω_j, we obtain the first Bianchi identity:

$$\Omega^i_j\wedge\omega^j=0,$$

or equivalently,

$$K^i_{jkl}+K^i_{klj}+K^i_{ljk}=0.$$

Hence, the condition $\sum K^i_{jil}=0$ implies also

$$\sum K^i_{ikl}=0.$$

We shall now prove the uniqueness of a normal Cartan connection. Let $\bar\omega=(\omega^i;\omega^i_j;\bar\omega)$ be another Cartan connection with the given $(\omega^i;\omega^i_j)$. By conditions (a) and (c) of § 3, we can write

$$\bar\omega_j-\omega_j=\sum A_{jk}\,\omega^k,$$

where the coefficients A_{jk} are functions on P. Denoting the curvature of $\bar\omega$ by $\bar\Omega=(0;\bar\Omega^i_j;\bar\Omega_j)$ and writing

$$\bar\Omega^i_j=\tfrac{1}{2}\sum\bar K^i_{jkl}\,\omega^k\wedge\omega^l,$$

we obtain by a straightforward computation using (II) the following relations between K^i_{jkl} and $\bar K^i_{jkl}$:

$$\bar K^i_{jkl}-K^i_{jkl}=-\delta^i_l A_{jk}+\delta^i_k A_{jl}+\delta^i_j A_{kl}-\delta^i_j A_{lk} \qquad \text{(projective)},$$

$$\bar K^i_{jkl}-K^i_{jkl}=-\delta^i_l A_{jk}+\delta^i_k A_{jl}+\delta^j_l A_{ik}-\delta^j_k A_{il}+\delta^i_j A_{kl}-\delta^i_j A_{lk} \qquad \text{(conformal)}.$$

Hence,

$$\left.\begin{aligned} &(\mathrm{i})_p \quad \sum (\bar{K}^i_{ikl} - K^i_{ikl}) = (n+1)(A_{kl} - A_{lk}), \\ &(\mathrm{ii})_p \quad \sum (\bar{K}^i_{jil} - K^i_{jil}) = (n-1) A_{jl} + (A_{jl} - A_{lj}) \end{aligned}\right\} \quad \text{(projective)}$$

$$\left.\begin{aligned} &(\mathrm{i})_c \quad \sum (\bar{K}^i_{jil} - K^i_{jil}) = (n-2) A_{jl} + \delta_{jl} \sum_i A_{ii}, \\ &(\mathrm{ii})_c \quad \sum_{i,j} (\bar{K}^i_{jij} - K^i_{jij}) = 2(n-1) \sum_i A_{ii}. \end{aligned}\right\} \quad \text{(conformal)}$$

If both ω and $\bar{\omega}$ are normal Cartan connections, i.e., $\sum K^i_{jil} = \sum \bar{K}^i_{jil} = 0$, then we see that $A_{ij} = 0$ and hence $\omega = \bar{\omega}$ in either the projective or the conformal case. This proves the uniqueness of a normal Cartan connection.

To prove the existence, assume first that there is a Cartan connection $\omega = (\omega^i; \omega^i_j; \omega_j)$ with the given $(\omega^i; \omega^i_j)$. We shall find a suitable system of functions A_{jk} so that $\bar{\omega} = (\omega^i; \omega^i_j; \bar{\omega}_j)$ becomes a normal Cartan connection. In the projective case, it is clear from $(\mathrm{i})_p$ and $(\mathrm{ii})_p$ that it suffices to set

$$A_{jk} = \frac{1}{(n+1)(n-1)} \sum K^i_{ijk} - \frac{1}{n-1} \sum K^i_{jik}.$$

In the conformal case, from $(\mathrm{i})_c$ and $(\mathrm{ii})_c$ we see that it suffices to set

$$A_{jk} = \frac{1}{n-2} \left\{ \frac{1}{2(n-1)} \delta_{jk} \sum K^i_{lil} - \sum K^i_{jik} \right\}.$$

To complete the proof of the theorem, we have now only to prove that there is at least one Cartan connection ω with the given $(\omega^i; \omega^i_j)$. Let $\{U_\alpha\}$ be a locally finite open cover of M with a partition of unity $\{f_\alpha\}$. If ω_α is a Cartan connection in $P \mid U_\alpha$ with the given $(\omega^i; \omega^i_j)$, then $\sum_\alpha (f_\alpha \circ \pi) \omega_\alpha$ is a Cartan connection in P with the given $(\omega^i; \omega^i_j)$, where $\pi: P \to M$ is the projection. Hence, the problem is reduced to the case where P is a product bundle. Fixing a cross section $\sigma: M \to P$, set $\omega_j(X) = 0$ for every vector X tangent to $\sigma(M)$. If Y is an arbitrary tangent vector of P, we can write uniquely

$$Y = R_a(X) + W,$$

where X is a vector tangent to $\sigma(M)$, a is in $\mathfrak{L}_0$ and W is a vertical vector. Extend W to a unique fundamental vector field A^* of P with $A \in \mathfrak{l}_0 = \mathfrak{g}_0 + \mathfrak{g}_1$. By conditions (a) and (b) for Cartan connections, we have to set

$$\omega(Y) = \operatorname{ad}(a^{-1})(\omega(X)) + A.$$

This defines the desired (ω_j). q.e.d.

Theorem 4.3. *Let P be as in Theorem 4.2 and $\omega=(\omega^i;\omega^i_j;\omega_j)$ be a normal Cartan connection. Then*

(1) $\sum \Omega^i_j \wedge \omega^j = 0$, *or equivalently,* $K^i_{jkl}+K^i_{klj}+K^i_{ljk}=0$, *where*

$$\Omega^i_j = \sum \tfrac{1}{2} K^i_{jkl}\, \omega^k \wedge \omega^l;$$

(2) $\sum \omega^i \wedge \Omega_i = 0$, *or equivalently,* $K_{jkl}+K_{klj}+K_{ljk}=0$, *where*

$$\Omega_j = \sum \tfrac{1}{2} K_{jkl}\, \omega^k \wedge \omega^l;$$

(3) *if* $\Omega^i_j=0$, *then* $\Omega_j=0$ *provided* $\dim M \geqq 3$ *in the projective case and* $\dim M \geqq 4$ *in the conformal case.*

Proof. (1) This has been already proved in the proof of Theorem 4.2 whether the connection is normal or not as long as it satisfies (I) of Theorem 4.2.

(2) Apply exterior differentiation d to $(\mathrm{II})_p$ and $(\mathrm{II})_c$, collect those terms involving only ω^i (not ω^i_j and ω_j) and take the trace. Since $d\Omega^i_i=0$, we obtain the desired result.

(3) Similarly, apply exterior differentiation d to $(\mathrm{II})_p$ and $(\mathrm{II})_c$ and collect those terms involving only ω^i. Since $\Omega^i_j=0$ by assumption, we obtain

$$\omega^i \wedge \Omega_j = 0 \qquad \text{(in the projective case)},$$

$$\omega^i \wedge \Omega_j - \omega^j \wedge \Omega_i = 0 \qquad \text{(in the conformal case)}.$$

In the projective case, we can conclude immediately that $\Omega_j=0$ provided $n \geqq 3$. In the conformal case, we rewrite the above identity as

$$\delta^i_m K_{jkl} - \delta^j_m K_{ikl} - \delta^i_k K_{jml} + \delta^j_k K_{iml} - \delta^i_l K_{jkm} + \delta^j_l K_{ikm} = 0.$$

Let $i=m$, and summing over i we obtain

$$(n-3) K_{jkl} + \delta^j_k \sum_i K_{iil} + \delta^j_l \sum_i K_{iki} = 0.$$

Summing over $j=k$, we obtain

$$(2n-4) \sum_i K_{iil} = 0.$$

From the last two identities we can conclude that $K_{jkl}=0$ provided $n \geqq 4$. q.e.d.

Theorem 4.2 goes back to E. Cartan [2, 4]. Ochiai [3] proved Theorem 4.2 in a very general setting, thus showing that the existence and uniqueness of a normal Cartan connection are related to the vanishing of certain Spencer cohomology groups. Here we followed Kobayashi-Nagano [1] rather closely. For a slightly different approach, see Tanaka [1, 2]. See also Ogiue [1].

5. Frames of Second Order

In preparation for the following section on projective and conformal structures, we shall construct bundles of frames of higher order contact, in particular, of second order contact (see § 8 of Chapter I).

Let M be an n-dimensional manifold. If U and V are two neighborhoods of the origin 0 of $\mathbf{R}^n$, two mappings $f\colon U \to M$ and $g\colon V \to M$ are said to define the same *r-jet* at 0 if they have the same partial derivatives up to order r at 0. The r-jet given by f is denoted by $j_0^r(f)$. If f is a diffeomorphism of a neighborhood of 0 onto an open subset of M, then the r-jet $j_0^r(f)$ at 0 is called an *r-frame* at $x=f(0)$. Clearly, a 1-frame is an ordinary linear frame. The set of r-frames of M, denoted by $P^r(M)$, is a principal bundle over M with natural projection π, $\pi(j_0^r(f))=f(0)$, and with structure group $G^r(n)$ which will be described next.

Let $G^r(n)$ be the set of r-frames $j_0^r(g)$ at $0 \in \mathbf{R}^n$, where g is a diffeomorphism from a neighborhood of 0 in $\mathbf{R}^n$ onto a neighborhood of 0 in $\mathbf{R}^n$. Then $G^r(n)$ is a group with multiplication defined by the composition of jets, i.e.,

$$j_0^r(g)\cdot j_0^r(g')=j_0^r(g\circ g').$$

The group $G^r(n)$ acts on $P^r(M)$ on the right by

$$j_0^r(f)\cdot j_0^r(g)=j_0^r(f\circ g) \quad \text{for } j_0^r(f)\in P^r(M) \quad \text{and} \quad j_0^r(g)\in G^r(n).$$

Clearly, $P^1(M)$ is the bundle of linear frames over M with group $G^1(n)=\mathrm{GL}(n;\mathbf{R})$.

From now on we shall consider only $P^1(M)$ and $P^2(M)$. If we consider the group $\mathfrak{A}(n;\mathbf{R})$ of affine transformations of $\mathbf{R}^n$ as a principal bundle over $\mathbf{R}^n=\mathfrak{A}(n;\mathbf{R})/\mathrm{GL}(n;\mathbf{R})$ with structure group $\mathrm{GL}(n;\mathbf{R})$, we have a natural bundle isomorphism between $\mathfrak{A}(n;\mathbf{R})$ and the bundle $P^1(\mathbf{R}^n)$ of linear frames over $\mathbf{R}^n$:

$$\begin{array}{ccc} \mathfrak{A}(n;\mathbf{R}) & \longleftrightarrow & P^1(\mathbf{R}^n) \\ \Big\downarrow & & \Big\downarrow \\ \mathbf{R}^n & \longleftrightarrow & \mathbf{R}^n \end{array}$$

Under this isomorphism, the identity e of $\mathfrak{A}(n;\mathbf{R})$ corresponds to $j_0^1(\mathrm{id})$, where id denotes the identity transformation of $\mathbf{R}^n$. We shall therefore denote $j_0^1(\mathrm{id})$ by e. The tangent space of $P^1(\mathbf{R}^n)$ at e will be identified with that of $\mathfrak{A}(n;\mathbf{R})$ at e, that is, with the Lie algebra

$$\mathfrak{a}(n;\mathbf{R})=\mathbf{R}^n+\mathfrak{gl}(n;\mathbf{R})$$

of $\mathfrak{A}(n;\mathbf{R})$.

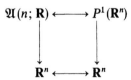

We shall now define a 1-form on $P^2(M)$ with values in $\mathfrak{a}(n; \mathbf{R})$. First, we observe that $j_0^2(f) \to j_0^1(f)$ defines a bundle homomorphism $P^2(M) \to P^1(M)$. Let X be a vector tangent to $P^2(M)$ at $j_0^2(f)$ and X' be the image of X under the homomorphism $P^2(M) \to P^1(M)$. Then X' is a vector tangent to $P^1(M)$ at $j_0^1(f)$. Since f is a diffeomorphism of a neighborhood of the origin 0 of $\mathbf{R}^n$ onto a neighborhood of $f(0) \in M$, it induces a diffeomorphism of a neighborhood of $e \in P^1(\mathbf{R}^n)$ onto a neighborhood of $j_0^1(f) \in P^1(M)$. The latter induces an isomorphism of the tangent space $\mathfrak{a}(n; \mathbf{R}) = T_e(P^1(\mathbf{R}^n))$ onto the tangent space of $P^1(M)$ at $j_0^1(f)$; this isomorphism will be denoted by $\tilde{f}$ and easily seen to depend only on $j_0^2(f)$. The *canonical form* θ on $P^2(M)$ is an $\mathfrak{a}(n; \mathbf{R})$-valued 1-form defined by

$$\theta(X) = \tilde{f}^{-1}(X').$$

The construction generalizes that of the canonical form of $P^1(M)$.

We define the *adjoint action* ad of $G^2(n)$ on $\mathfrak{a}(n; \mathbf{R})$ as follows. Let $j_0^2(g) \in G^2(n)$ and $j_0^1(f) \in P^1(\mathbf{R}^n)$. The mapping of a neighborhood of $e \in P^1(\mathbf{R}^n)$ onto a neighborhood of $e \in P^1(\mathbf{R}^n)$ defined by

$$j_0^1(f) \to j_0^1(g \circ f \circ g^{-1})$$

induces a linear isomorphism of the tangent space $\mathfrak{a}(n; \mathbf{R}) = T_e(P^1(\mathbf{R}^n))$ onto itself. This linear automorphism of $\mathfrak{a}(n; \mathbf{R})$ depends only on $j_0^2(g)$ and will be denoted by $\mathrm{ad}(j_0^2(g))$.

Since $G^2(n)$ acts on $P^2(M)$ on the right, every element A of the Lie algebra $\mathfrak{g}^2(n)$ of $G^2(n)$ induces a vector field A^* on $P^2(M)$, called the *fundamental vector field* corresponding to A

Proposition 5.1. *Let θ be the canonical form on $P^2(M)$ defined above. Then*

(1) $\theta(A^*) = A'$ for $A \in \mathfrak{g}^2(n)$, where $A' \in \mathfrak{g}^1(n) = \mathfrak{gl}(n; \mathbf{R})$ is the image of A under the homomorphism $\mathfrak{g}^2(n) \to \mathfrak{g}^1(n)$;

(2) $(R_a)^* \theta = \mathrm{ad}(a^{-1})\theta$ for $a \in G^2(n)$.

The proof is straightforward.

We shall now express the canonical form of $P^2(M)$ in terms of the local coordinate system of $P^2(M)$ which arises in a natural way from a local coordinate system of M. For this purpose we may restrict ourselves to the case $M = \mathbf{R}^n$. Let $e_1, \ldots, e_n$ be the natural basis for $\mathbf{R}^n$ and $(x^1, \ldots, x^n)$ be the natural coordinate system in $\mathbf{R}^n$. Each 2-frame u of $\mathbf{R}^n$ has a unique polynomial representation $u = j_0^2(f)$ of the form

$$f(x) = \sum (u^i + \sum u^i_j x^j + \tfrac{1}{2} \sum u^i_{jk} x^j x^k)\, e_i,$$

where $x = \sum x^i e_i$ and $u^i_{jk} = u^i_{kj}$. We take $(u^i; u^i_j; u^i_{jk})$ as the *natural coordinate system in* $P^2(\mathbf{R}^n)$. Restricting $(u^i_j; u^i_{jk})$ to $G^2(n)$ we obtain the *natural coordinate system in* $G^2(n)$, which will be denoted by $(s^i_j;\, s^i_{jk})$. The action of $G^2(n)$ on $P^2(\mathbf{R}^n)$ is then given by

$$(u^i;\, u^i_j;\, u^i_{jk})(s^i_j;\, s^i_{jk}) = (u^i;\, \sum u^i_p s^p_j;\, \sum u^i_p s^p_{jk} + \sum u^i_{qr} s^q_j s^r_k).$$

In particular, the multiplication in $G^2(n)$ is given by

$$(\bar{s}^i_j;\, \bar{s}^i_{jk})(s^i_j;\, s^i_{jk}) = (\sum \bar{s}^i_p s^p_j;\, \sum \bar{s}^i_p s^p_{jk} + \sum \bar{s}^i_{qr} s^q_j s^r_k).$$

Similarly, we can introduce the natural coordinate system $(u^i;\, u^i_j)$ in $P^1(\mathbf{R}^n)$ and the natural coordinate system (s^i_j) in $G^1(n)$ so that the homomorphisms $P^2(\mathbf{R}^n) \to P^1(\mathbf{R}^n)$ and $G^2(n) \to G^1(n)$ are given by

$$(u^i;\, u^i_j;\, u^i_{jk}) \to (u^i;\, u^i_j) \quad \text{and} \quad (s^i_j;\, s^i_{jk}) \to (s^i_j),$$

respectively. Let $\{E_i;\, E^j_i\}$ be the basis for $\mathfrak{a}(n;\mathbf{R})$ defined by

$$E_i = (\partial/\partial u^i)_e, \qquad E^j_i = (\partial/\partial u^i_j)_e$$

and set

$$\theta = \sum \theta^i E_i + \sum \theta^i_j E^j_i.$$

From the definition of the canonical form θ and the action of $G^2(n)$ on $P^2(\mathbf{R}^n)$ expressed in terms of the natural coordinate systems, we obtain the following formulae:

$$d u^i = \sum u^i_j \theta^j,$$
$$d u^i_j = \sum u^i_k \theta^k_j + \sum u^i_{hj} \theta^h.$$

Let (v^i_j) be the inverse matrix of (u^i_j). Then

$$\theta^i = \sum v^i_k\, d u^k,$$
$$\theta^i_j = \sum v^i_k\, d u^k_j - \sum v^i_k u^k_{hj} v^h_l\, d u^l.$$

From these formulae we obtain the following structure equation:

Proposition 5.2. *Let* $\theta = (\theta^i;\, \theta^i_j)$ *be the canonical form on* $P^2(M)$. *Then*

$$d\theta^i = -\sum \theta^i_k \wedge \theta^k.$$

For the canonical form on $P^r(M)$ for $r > 2$ and its structure equation, see Kobayashi [8].

This set-up of jets, frames of higher order contact, etc. is due to Ehresmann [3].

6. Projective and Conformal Structures

Let $\mathfrak{L}/\mathfrak{L}_0$ be as in Example 4.1 or 4.2. We can consider $\mathfrak{L}_0$ as a subgroup of the group $G^2(n) = \{(a^i_j;\, a^i_{jk})\}$ defined in § 5 as follows. Let o denote the origin of the homogeneous space $\mathfrak{L}/\mathfrak{L}_0$ and consider each element g of

$\mathfrak{L}_0$ as a transformation of $\mathfrak{L}/\mathfrak{L}_0$ leaving the origin o fixed. It can be easily verified that $j_0^2(g)=\text{id}$ if and only if $g=\text{id}$, that is, every element of $\mathfrak{L}_0$ is determined by its partial derivatives of order 1 and 2 at the origin o. Hence, $\mathfrak{L}_0$ is isomorphic to the group of 2-jets $\{j_0^2(g);\, g\in\mathfrak{L}_0\}$. Choosing a basis for $\mathfrak{g}_{-1}$, identify $\mathfrak{g}_{-1}$ with $V=\mathbf{R}^n$ as in Example 4.1 or 4.2. Then the mapping

$$\mathbf{R}^n=\mathfrak{g}_{-1}\xrightarrow{\exp}\mathfrak{L}\to\mathfrak{L}/\mathfrak{L}_0$$

gives a diffeomorphism from a neighborhood of $0\in\mathbf{R}^n$ onto a neighborhood of $o\in\mathfrak{L}/\mathfrak{L}_0$ and defines a local coordinate system around $o\in\mathfrak{L}/\mathfrak{L}_0$. With respect to this coordinate system, each 2-jet $j_0^2(g)$ is an element of the group $G^2(n)$. Hence, $\mathfrak{L}_0$ can be considered as a subgroup of $G^2(n)$. An explicit description of $\mathfrak{L}_0$ as a subgroup of $G^2(n)$ is not without interest although it is not essential in the subsequent discussion. Let $N^2(n)=\{(a^i_{jk})\}$ denote the kernel of the natural homomorphism $G^2(n)\to G^1(n)$ which sends $(a^i_j;\, a^i_{jk})$ into (a^i_j). In view of the diagram of exact sequences:

$$\begin{array}{ccccccccc} 0 & \to & N^2(n) & \to & G^2(n) & \to & G^1(n) & \to & 1 \\ & & \cup & & \cup & & \cup & & \\ 0 & \to & \mathfrak{L}_1 & \to & \mathfrak{L}_0 & \to & \mathfrak{L}_0/\mathfrak{L}_1 & \to & 1, \end{array}$$

we shall describe $\mathfrak{L}_0/\mathfrak{L}_1$ as a subgroup of $G^1(n)$ and $\mathfrak{L}_1$ as a subgroup of $N^2(n)$.

First, let $\mathfrak{L}/\mathfrak{L}_0$ be as in Example 4.1. Then $\mathfrak{L}_0/\mathfrak{L}_1=G^1(n)=GL(n;\mathbf{R})$ and $\mathfrak{L}_1=\{(a^i_{jk});\, a^i_{jk}=\delta^i_j\xi_k+\delta^i_k\xi_j\}$.

Next, let $\mathfrak{L}/\mathfrak{L}_0$ be as in Example 4.2. Then $\mathfrak{L}_0/\mathfrak{L}_1=CO(n)$ and

$$\mathfrak{L}_1=\{(a^i_{jk});\, a^i_{jk}=\delta^i_j\xi_k+\delta^i_k\xi_j-\delta^j_k\xi_i\}.$$

Thus $\mathfrak{L}_1$ coincides with the first prolongation of the Lie algebra $\mathfrak{co}(n)$ (see Example 2.6 of Chapter I).

Let M be a manifold of dimension n and $P^2(M)$ be the bundle of 2-frames over M with structure group $G^2(n)$ (see § 5). A principal subbundle P of $P^2(M)$ with structure group $\mathfrak{L}_0$ ($\subset G^2(n)$) is called a *projective structure* or a *conformal structure* on M according as $\mathfrak{L}/\mathfrak{L}_0$ is as in Example 4.1 or 4.2. (In Example 2.6 of Chapter I, a $CO(n)$-structure on M was called a conformal structure on M. The two definitions are equivalent in the following sense. If P is a conformal structure as a subbundle of $P^2(M)$, then $P/\mathfrak{L}_1$ is a subbundle of $P^1(M)=L(M)$ with group $\mathfrak{L}_0/\mathfrak{L}_1=CO(n)$ in a natural manner and hence is a $CO(n)$-structure. Conversely, if Q is a $CO(n)$-structure on M, then its first prolongation Q_1 (see § 5 of Chapter I) is a conformal structure as a subbundle of $P^2(M)$. This gives a one-to-one correspondence between the conformal structures $P\subset P^2(M)$ and the $CO(n)$-structures $Q\subset P^1(M)$. Since we shall not use this fact, the proof is left to the reader. But a projective structure P cannot be defined as a subbundle of $P^1(M)=L(M)$ since $P/\mathfrak{L}_1$ is $P^1(M)$ itself and the first prolongation of $P^1(M)$ is precisely $P^2(M)$.)

Now, let $P\subset P^2(M)$ be a projective or conformal structure on M and $(\omega^i;\omega^i_j)$ be the restriction to P of the canonical form $(\theta^i;\theta^i_j)$ of $P^2(M)$ (see § 5). By Propositions 5.1 and 5.2, $(\omega^i;\omega^i_j)$ satisfies the assumptions (a′), (b′), (c′) and (I) of Theorem 4.2. We shall therefore call $(\omega^i;\omega^i_j)$ the *canonical form* of P. By Theorem 4.2, there is a unique normal projective or conformal connection $(\omega^i;\omega^i_j;\omega_j)$ provided $n\geqq 3$.

Let P and P' be projective (resp. conformal) structures on manifolds M and M', respectively. A (local) diffeomorphism f of M into M' induces a local isomorphism f_* of the bundle $P^2(M)$ of 2-frames of M into the bundle $P^2(M')$ of 2-frames of M'. If f_* sends P into P', then f is called a (local) *projective* (resp. *conformal*) *isomorphism* of M into M'. If $M=M'$ and $P=P'$, then a projective (resp. conformal) isomorphism f is called a *projective* (resp. *conformal*) *transformation* or *automorphism*. This definition is completely analogous to that of an automorphism of a G-structure given in § 1 of Chapter I. An infinitesimal projective or conformal transformation can be defined in the same way as an infinitesimal automorphism of a G-structure.

Assume that $n\geqq 3$ so that the normal projective or conformal connection is unique. Then, for each automorphism f of P, f_* (restricted to P) preserves the normal connection $\omega=(\omega^i;\omega^i_j;\omega_j)$. From Theorem 3.1 we obtain (Kobayashi [2]).

Theorem 6.1. *If P is a projective or conformal structure on a manifold M of dimension $n\geqq 3$, then the group $\mathfrak{A}$ of projective or conformal transformations is a Lie transformation group of dimension $\leqq \dim P$ $\left(=n^2+3n\right.$ in the projective case and $\left.=\frac{1}{2}(n+1)(n+2)\right.$ in the conformal case$\left.\right)$.*

In order to determine the cases where $\dim\mathfrak{A}=\dim P$, we consider some examples.

Let $\mathfrak{L}/\mathfrak{L}_0=P_n(\mathbf{R})$ be as in Example 4.1, where $\mathfrak{L}=\mathrm{PGL}(n;\mathbf{R})$. The principal bundle $\mathfrak{L}$ over $\mathfrak{L}/\mathfrak{L}_0$ with group $\mathfrak{L}_0$ can be identified with a projective structure on $\mathfrak{L}/\mathfrak{L}_0$ in a natural manner. To describe this identification, let o denote the origin (i.e., the coset $\mathfrak{L}_0$) of the homogeneous space $\mathfrak{L}/\mathfrak{L}_0$. Since each $f\in\mathfrak{L}$ is a transformation of $\mathfrak{L}/\mathfrak{L}_0$ and a neighborhood of o in $\mathfrak{L}/\mathfrak{L}_0$ is identified with a neighborhood of 0 in $\mathbf{R}^n$ in a natural way, the 2-jet $j^2_0(f)$ can be considered as a 2-frame of $\mathfrak{L}/\mathfrak{L}_0$ at $f(\mathrm{o})$. The set of all 2-frames thus obtained defines a projective structure on $\mathfrak{L}/\mathfrak{L}_0$ which can be identified with the bundle $\mathfrak{L}$ over $\mathfrak{L}/\mathfrak{L}_0$. Then the Maurer-Cartan form $\omega=(\omega^i;\omega^i_j;\omega_j)$ of $\mathfrak{L}$ defined in Example 4.1 defines the normal projective connection of this projective structure. The Maurer-Cartan structure equations of $\mathfrak{L}$ show that the connection has no curvature. Clearly, $\mathfrak{L}$ is the group of projective transformations of this projective structure.

We consider now the universal covering space S^n of $P_n(\mathbf{R})$. Since the covering projection $S^n \to P_n(\mathbf{R})$ is a local diffeomorphism, the natural projective structure on $P_n(\mathbf{R})$ described above induces a projective structure on S^n. We shall give a group-theoretic description of this natural projective structure on S^n. Let $\mathfrak{L} = \mathrm{GL}(n+1; \mathbf{R})/\mathbf{R}^+$, where $\mathbf{R}^+$ denotes the normal subgroup of $\mathrm{GL}(n+1; \mathbf{R})$ consisting of elements aI_{n+1} with $a>0$. It is a group with two components which is locally isomorphic to $\mathrm{PLG}(n; \mathbf{R})$. ($\mathrm{PLG}(n; \mathbf{R}) = \mathrm{GL}(n+1; \mathbf{R})/\mathbf{R}^*$, where $\mathbf{R}^*$ is the normal subgroup consisting of elements aI_{n+1} with $a \neq 0$.) Let

$$\mathfrak{L}_0 = \left\{ \begin{pmatrix} A & 0 \\ \xi & a \end{pmatrix} \in \mathrm{GL}(n+1; \mathbf{R});\ a>0 \right\} \Big/ \mathbf{R}^+,$$

where $A \in \mathrm{GL}(n; \mathbf{R})$ and ξ is a row n-vector. Then $S^n = \mathfrak{L}/\mathfrak{L}_0$. The principal bundle $\mathfrak{L}$ over $\mathfrak{L}/\mathfrak{L}_0$ with group $\mathfrak{L}_0$ is the desired projective structure on S^n which is locally isomorphic to the natural projective structure on $P_n(\mathbf{R})$ under the covering projection $S^n \to P_n(\mathbf{R})$. The group of projective transformations of this projective structure is $\mathfrak{L}$.

As the third example, let $\mathfrak{L}/\mathfrak{L}_0 = S^n$ be as in Example 4.2, where $\mathfrak{L} = \mathrm{O}(n+1, 1)$. Then the principal bundle $\mathfrak{L}$ over $\mathfrak{L}/\mathfrak{L}_0$ can be naturally identified with a conformal structure on $\mathfrak{L}/\mathfrak{L}_0$. Again the Maurer-Cartan form $\omega = (\omega^i; \omega^i_j; \omega_j)$ of $\mathfrak{L}$ defined in Example 4.2 defines the normal conformal connection of this conformal structure. As we can see from the Maurer-Cartan structure equations of $\mathfrak{L}$, the connection has no curvature. The group $\mathfrak{L}$ is precisely the group of conformal transformations of this conformal structure.

Theorem 6.2. *In Theorem* 6.1, *assume* $\dim \mathfrak{A} = \dim P$. *In the projective case, P is the natural projective structure on either $P_n(\mathbf{R})$ or its universal covering space S^n as explained above. In the conformal case, P is the natural conformal structure on S^n described above.*

We shall only indicate the main idea of the proof. We consider the projective case, the conformal case being similar. Since each 1-parameter group of projective transformations of M lifts to a 1-parameter group of projective transformations of the universal covering space $\tilde{M}$, the group of projective transformations of $\tilde{M}$ has the maximum dimension $\dim P$. We shall first assume that M is simply connected. Choose a point u_0 of P. We know (Theorem 3.2 of Chapter I) that the mapping $f \in \mathfrak{A} \to f_*(u_0) \in P$ is injective and the image is a closed submanifold of P which can be identified with $\mathfrak{A}$ (as a differentiable manifold). Since $\dim \mathfrak{A} = \dim P$ by assumption, the identity component of $\mathfrak{A}$ can be identified with one of the components of P under the mapping $\mathfrak{A} \to P$ defined above. (Since P

has two connected components, $\mathfrak{A}$ can be identified with P if $\mathfrak{A}$ is not connected.) Let x_0 be the base point of u_0 in M and $\mathfrak{A}_0$ be the isotropy subgroup of $\mathfrak{A}$ at x_0. Since $\mathfrak{A}$ is fibre-transitive on P, it is transitive on M so that $M=\mathfrak{A}/\mathfrak{A}_0$. Let $\omega=(\omega^i;\omega^i_j;\omega_j)$ be the normal projective connection of P. Since the forms $(\omega^i;\omega^i_j;\omega_j)$ are invariant by $\mathfrak{A}$, restricted to $\mathfrak{A}\subset P$ they are Maurer-Cartan forms of the group $\mathfrak{A}$. Since the curvature of ω vanishes by Theorem 3.2, the structure equations (I), $(\mathrm{II})_p$, (III) of § 5 are nothing but the Maurer-Cartan structure equations of the group $\mathfrak{A}$. From these structure equations we see that $\mathfrak{A}$ is locally isomorphic to $\mathfrak{L}=\mathrm{GL}(n+1;\mathbf{R})/\mathbf{R}^+$. The identity component of $\mathfrak{A}_0$ coincides with the identity component of the structure group $\mathfrak{L}_0$. Since both $\mathfrak{A}/\mathfrak{A}_0$ and $\mathfrak{L}/\mathfrak{L}_0$ are simply connected, the standard argument proves that the identity component of $\mathfrak{A}$ and the identity component of $\mathfrak{L}$ are isomorphic to each other not only as groups but also as bundles over $M=\mathfrak{A}/\mathfrak{A}_0$ and $S^n=\mathfrak{L}/\mathfrak{L}_0$ respectively. Then it follows that the bundle P over M is isomorphic to the bundle $\mathfrak{L}$ over S^n. If M is not simply connected, then $M=S^n/\Gamma$, where Γ is a discrete subgroup of $\mathfrak{L}=\mathrm{GL}(n+1;\mathbf{R})/\mathbf{R}^+$ which commutes elementwise with the identity component $\mathrm{GL}^+(n+1;\mathbf{R})/\mathbf{R}^+$ of $\mathfrak{L}$ as in the proof of Theorem 3.1 of Chapter II. (Here $\mathrm{GL}^+(n+1;\mathbf{R})$ denotes the identity component of $\mathrm{GL}(n+1;\mathbf{R})$, which consists of matrices with positive determinant.) Then we see that Γ consists of two elements represented by $\pm I_{n+1}$, and that P is the natural projective structure on $M=P_n(\mathbf{R})$.

7. Projective and Conformal Equivalences

We shall first explain projective equivalence of affine connections. Let $\mathfrak{l}=\mathfrak{g}_{-1}+\mathfrak{g}_0+\mathfrak{g}_1=V+\mathfrak{gl}(n;\mathbf{R})+V^*$ be as in Example 4.1. Let $P^1(M)$ $(=L(M))$ be the bundle of linear frames over M and $\theta=(\theta^i)$ be the canonical form of $P^1(M)$, viewed as a $\mathfrak{g}_{-1}$-valued 1-form as well as a V-valued 1-form. Two torsionfree affine connections of M defined by $\mathfrak{g}_0$-valued 1-forms $\omega=(\omega^i_j)$ and $\omega'=(\omega'^i_j)$ are said to be *projectively equivalent* if there exists a $\mathfrak{g}_1$-valued function $p=(p_j)$ on $P^1(M)$ such that

$$\omega'-\omega=[\theta,p].$$

This formulation is due to Tanaka [4] (see also Kobayashi-Ochiai [1]). Note that the left hand side takes values in $\mathfrak{g}_0$ and the right hand side in $[\mathfrak{g}_{-1},\mathfrak{g}_1]\subset\mathfrak{g}_0$. If we consider ω and ω' as $\mathfrak{gl}(n;\mathbf{R})$-valued forms, θ as a V-valued form (i.e., a form whose values are column n-vectors) and p as a V^*-valued function (i.e., a function whose values are row n-vectors), then the relation above may be rewritten as follows:

$$\omega'-\omega=\theta\,p+(p\,\theta)\,I_n,$$

or more explicitly

$$\omega_j'^i - \omega_j^i = \theta^i p_j + (\sum_k \theta^k p_k)\, \delta_j^i.$$

If we set

$$\omega_j'^i - \omega_j^i = \sum_k r_{jk}^i \theta^k,$$

then the relation reads as follows:

$$r_{jk}^i = \delta_k^i p_j + \delta_j^i p_k.$$

Similarly, we define conformal equivalence of affine connections. Let $\mathfrak{l} = \mathfrak{g}_{-1} + \mathfrak{g}_0 + \mathfrak{g}_1 = V + \mathfrak{co}(n) + V^*$ be as in Example 4.2. Let $Q \subset P^1(M)$ be a $CO(n)$-structure over M and $\theta = (\theta^i)$ be the canonical form on Q. Two torsionfree connections in Q defined by $\mathfrak{g}_0$-valued 1-forms $\omega = (\omega_j^i)$ and $\omega' = (\omega_j'^i)$ on Q are said to be *conformally equivalent* if there exists a $\mathfrak{g}_1$-valued function $p = (p_j)$ on Q such that

$$\omega' - \omega = [\theta, p],$$

or, in terms of matrix and vector notations,

$$\omega' - \omega = \theta\, p - {}^t p\, {}^t\theta + (p\,\theta)\, I_n,$$

or more explicitly,

$$\omega_j'^i - \omega_j^i = \theta^i p_j - \theta^j p_i + (\sum_k \theta^k p_k)\, \delta_j^i.$$

If we set

$$\omega_j'^i - \omega_j^i = \sum_k r_{jk}^i \theta^k,$$

then the relation reads as follows:

$$r_{jk}^i = \delta_k^i p_j - \delta_k^j p_i + \delta_j^i p_k.$$

Although two torsionfree connections in $P^1(M)$ are not necessarily projectively equivalent, two torsionfree connections $\omega = (\omega_j^i)$ and $\omega' = (\omega_j'^i)$ in the $CO(n)$-structure Q are always conformally equivalent to each other. This difference comes from the fact that while $\mathfrak{g}_1$ is the first prolongation of $\mathfrak{g}_0$ in the conformal case, $\mathfrak{g}_1$ is strictly smaller than the first prolongation of $\mathfrak{g}_0$ in the projective case. In fact, at each point of Q, (r_{jk}^i) defines an element of the first prolongation of $\mathfrak{g}_0$ since, for each fixed k, the matrix $(r_{jk}^i)_{i,j=1,\dots,n}$ is in $\mathfrak{co}(n)$ and (r_{jk}^i) is symmetric in j and k. In Example 2.6 of Chapter I, we proved that each element (r_{jk}^i) of the first prolongation of $\mathfrak{co}(n)$ determines a unique vector (p_j) such that $r_{jk}^i = \delta_k^i p_j - \delta_k^j p_i + \delta_j^i p_k$.

In order to relate the notion of projective equivalence to that of projective structure, we prove the following

Proposition 7.1. *Let $P^2(M)$ be the bundle of 2-frames over an n-dimensional manifold M with structure group $G^2(n)$. Consider $G^1(n)=\mathrm{GL}(n;\mathbf{R})$ as the subgroup of $G^2(n)$ consisting of elements $(s^i_j;\, s^i_{jk})$ with $s^i_{jk}=0$ in terms of the natural coordinate system introduced in §5. Let $\mathfrak{L}/\mathfrak{L}_0=P_n(\mathbf{R})$ be as in Example 4.1 and consider $\mathfrak{L}_0$ as a subgroup of $G^2(n)$ as in §6 so that $G^1(n)\subset\mathfrak{L}_0\subset G^2(n)$. Then*

(1) *The cross sections $M\to P^2(M)/G^1(n)$ are in one-to-one correspondence with the torsionfree affine connections of M.*

(2) *The cross sections $M\to P^2(M)/\mathfrak{L}_0$ are in one-to-one correspondence with the projective structures of M.*

Proof. (1) Let $(u^i;\, u^i_j;\, u^i_{jk})$ be the natural local coordinate system in $P^2(M)$ induced from a local coordinate system $x^1,\ldots,x^n$ of M as in §5. We introduce a local coordinate system $(z^i;\, z^i_{jk})$ in $P^2(M)/G^1(n)$ in such a way that the natural projection $P^2(M)\to P^2(M)/G^1(n)$ is given by the equations

$$z^i=u^i,\qquad z^i_{jk}=\sum u^i_{pq}\,v^p_j\,v^q_k,$$

where

$$(v^i_j)=(u^i_j)^{-1}\qquad \text{(the inverse matrix)}.$$

Then a cross section $\Gamma\colon M\to P^2(M)/G^1(n)$ is given locally by a set of functions

$$z^i_{jk}=-\Gamma^i_{jk}(x^1,\ldots,x^n)\qquad\text{with } \Gamma^i_{jk}=\Gamma^i_{kj}.$$

If we consider the action of the group $G^2(n)$ on the fibre $G^2(n)/G^1(n)$, then we see that the functions Γ^i_{jk} behave under the coordinate changes as Christoffel's symbols should. This proves (1).

(2) Since the reductions of the structure group $G^2(n)$ to $\mathfrak{L}_0$ are in one-to-one correspondence with the cross sections $M\to P^2(M)/\mathfrak{L}_0$, (2) is evident. q.e.d.

Let $\theta=(\theta^i;\,\theta^i_j)$ be the canonical form on $P^2(M)$ defined in §5 and $\gamma\colon P^1(M)\hookrightarrow P^2(M)$ be the reduction of the structure group $G^2(n)$ to $G^1(n)$ corresponding to a cross section $\Gamma\colon M\to P^2(M)/G^1(n)$. Then $(\gamma^*(\theta^i))$ is the canonical form of $P^1(M)$ and $(\gamma^*(\theta^i_j))$ is the connection form corresponding to Γ. This follows easily from the proof of Proposition 7.1 and the expression of θ given in §5 in terms of the coordinate system $(u^i;\, u^i_j;\, u^i_{jk})$.

Every torsionfree affine connection $\Gamma\colon M\to P^2(M)/G^1(n)$, composed with the natural mapping $P^2(M)/G^1(n)\to P^2(M)/\mathfrak{L}_0$, gives a projective structure $M\to P^2(M)/\mathfrak{L}_0$. A torsionfree connection Γ is said to *belong to* a projective structure P if it induces P in the way described above.

Proposition 7.2. *Two torsionfree affine connections of a manifold M are projectively equivalent if and only if they belong to the same projective structure.*

Proof. Let Γ and Γ' be two cross sections $M \to P^2(M)/G^1(n)$, i.e., torsionfree affine connections and let ω and ω' be the corresponding connection forms on $P^1(M)$. A straightforward calculation shows that ω and ω' are projectively equivalent if and only if there is a 1-form $p = \sum p_i dx^i$ on M such that

$$\Gamma'^i_{jk} - \Gamma^i_{jk} = \delta^i_j p_k + \delta^i_k p_j,$$

where Γ'^i_{jk} and Γ^i_{jk} are Christoffel's symbols for ω and ω' with respect to a local coordinate system $x^1, \ldots, x^n$. This in turn is equivalent to the condition that Γ and Γ' induce the same cross section $M \to P^2(M)/\mathfrak{L}_0$, because the kernel $\mathfrak{L}_1$ of the homomorphism $\mathfrak{L}_0 \to G^1(n)$ consists of elements (a^i_{jk}) of the form $a^i_{jk} = \delta^i_j p_k + \delta^i_k p_j$ (see § 6). q.e.d.

Given a torsionfree affine connection on M, let $\mathfrak{A}(M)$ be the group of affine transformations of M and $\mathfrak{P}(M)$ be the group of projective transformations, i.e., automorphisms of the induced projective structure. Proposition 7.2 implies that a transformation of M is a projective transformation if and only if it transforms the given connection into a projectively equivalent affine connection. Evidently, we have the inclusion $\mathfrak{A}(M) \subset \mathfrak{P}(M)$. There are cases where these groups have the same identity component. We quote only the following result of Nagano [8].

Theorem 7.3. *Let M be a complete Riemannian manifold with parallel Ricci tensor. Then the largest connected group $\mathfrak{P}^0(M)$ of projective transformations of M coincides with the largest connected group $\mathfrak{A}^0(M)$ of affine transformations of M unless M is a space of positive constant curvature.*

For other related results, see Couty [2, 3], Ishihara [2, 3], Solodovnikov [1], Tanaka [2], Tashiro [3], Yano [3], Yano-Nagano [2], and references therein.

Given a $CO(n)$-structure $Q \subset P^1(M)$, let $P \subset P^2(M)$ be the corresponding conformal structure on M. Then a transformation of M is an automorphism of the $CO(n)$-structure Q if and only if it is an automorphism of the conformal structure P. Such a transformation is called a *conformal transformation* of M. If M is a Riemannian manifold, then a transformation of M is a conformal transformation with respect to the naturally induced $CO(n)$-structure if and only if it sends the metric into a conformally equivalent metric (cf. Example 2.6 of Chapter I). Evidently, the group $\mathfrak{C}(M)$ of conformal transformations of M contains the group $\mathfrak{I}(M)$ of isometries of M. We quote only two results on the relationship between $\mathfrak{C}(M)$ and $\mathfrak{I}(M)$.

Theorem 7.4. *Let M be a complete Riemannian manifold with parallel Ricci tensor. Then the largest connected group $\mathfrak{C}^0(M)$ of conformal trans-*

formations of M coincides with the largest connected group $\mathfrak{I}^0(M)$ of isometries of M unless M is isometric to a simply connected space of positive constant curvature (i.e., a sphere).

This result of Nagano [1] extends the result of Yano -Nagano [1] on Einstein manifolds. The following theorem was proved by Obata [8]. See Lelong-Ferrand [4] for related results.

Theorem 7.5. *Let M be a compact Riemannian manifold with constant scalar curvature. Then $\mathfrak{C}^0(M)=\mathfrak{I}^0(M)$ unless M is isometric to a simply connected space of positive constant curvature.*

For other related results, see Goldberg-Kobayashi [1], Kulkarni [1], Ledger-Obata [1], Lichnerowicz [1], Obata [3–7, 9], Suyama-Tsukamoto [1], Tanaka [1], Tashiro [2], Tashiro-Miyashita [1, 2], Weber-Goldberg [1], Yano [3, 8], Yano-Nagano [2], Yano-Obata [1], and references therein. A list of papers which had given partial results toward Theorem 7.5 can be found in Yano-Obata [1].

Appendices

1. Reductions of 1-Forms and Closed 2-Forms

For the discussion of contact structures and symplectic structures, it is important to know the simplest possible expressions for 1-forms and closed 2-forms.

Let ω be a 1-form defined on a manifold M of dimension n. We set

$$\omega^{(2k)} = d\omega \wedge \cdots \wedge d\omega \qquad (k \text{ times}),$$
$$\omega^{(2k+1)} = \omega \wedge d\omega \wedge \cdots \wedge d\omega \qquad (d\omega\text{: } k \text{ times}).$$

Then $\omega^{(p)}$ is a form of degree p. We say that a 1-form ω is of rank r at a point o if $\omega^{(r)} \neq 0$ but $\omega^{(r+1)} = 0$ at o. Then we have

Theorem 1. *If a 1-form ω is of rank p in a neighborhood of a point* o, *then there exists a local coordinate system $x^1, \ldots, x^n$ around* o *such that*

$$\omega = x^1\, dx^2 + x^3\, dx^4 + \cdots + x^{2k-1}\, dx^{2k} \qquad \textit{for } p = 2k,$$
$$\omega = x^1\, dx^2 + x^3\, dx^4 + \cdots + x^{2k-1}\, dx^{2k} + dx^{2k+1} \qquad \textit{for } p = 2k+1.$$

Proof. We first prove

Lemma 1. *If ω is of rank p, then there exist a neighborhood U^n of* o, *a fibring π: $U^n \to U^p$ with* $\dim U^p = p$ *and a 1-form $\tilde{\omega}$ on U^p such that $\pi^*(\tilde{\omega}) = \omega$. In other words, with a suitable choice of local coordinate system, ω depends only on the first p coordinate functions.*

Proof of Lemma 1. At each point we consider the space S of tangent vectors X such that

$$\iota_X \omega = 0 \quad \text{and} \quad \iota_X d\omega = 0,$$

where ι_X denotes the interior product. It is straightforward to verify that if X and Y are vector fields which belong to S at each point, then $[X, Y]$ belongs to S at each point. We shall show that $\dim S = n - p$. (That will mean that S defines an involutive distribution of dimension $n-p$.) Set $p = 2k$ or $p = 2k+1$ according as p is even or odd. Then at each point,

$d\omega$ is a skew-symmetric bilinear form of rank $2k$ on the tangent space. If we denote by S' the space of tangent vectors X such that

$$\iota_X \, d\omega = 0,$$

then $\dim S' = n - 2k$ and $S \subset S'$. If $p = 2k$, then $S = S'$. In fact, if $S \neq S'$ and $Y \in S'$ is an element not in S, then $\omega(Y) \neq 0$. If $X_1, \ldots, X_{2k}$ are tangent vectors (not in S') such that

$$\omega^{(2k)}(X_1, \ldots, X_{2k}) = (d\omega \wedge \cdots \wedge d\omega)(X_1, \ldots, X_{2k}) \neq 0,$$

then

$$\begin{aligned}\omega^{(2k+1)}(Y, X_1, \ldots, X_{2k}) &= (\omega \wedge d\omega \wedge \cdots \wedge d\omega)(Y, X_1, \ldots, X_{2k}) \\ &= \omega(Y) \cdot \omega^{(2k)}(X_1, \ldots, X_{2k}) \neq 0,\end{aligned}$$

in contradiction to the assumption that $\omega^{(2k+1)} = 0$. If $p = 2k+1$, then $\dim S' = 1 + \dim S$. In fact, since $S = \{X \in S';\ \omega(X) = 0\}$, we have $\dim S \leqq \dim S' \leqq 1 + \dim S$. It suffices therefore to prove $S \neq S'$. Since $\omega^{(2k+1)} \neq 0$, there exists a $(2k+1)$-dimensional subspace T of the tangent space at each point such that the restriction of $\omega^{(2k+1)}$ to T is nonzero. In other words, for any basis $X_1, \ldots, X_{2k+1}$ of T, we have $\omega^{(2k+1)}(X_1, \ldots, X_{2k+1}) \neq 0$. But, if $S = S'$, then

$$\dim(S \cap T) \geqq \dim S + \dim T - n = n - 2k + 2k + 1 - n = 1.$$

By taking a basis $X_1, \ldots, X_{2k+1}$ of T in such a way that $X_1 \in S \cap T$, we obtain

$$\begin{aligned}\omega^{(2k+1)}(X_1, X_2, \ldots, X_{2k+1}) &= (\omega \wedge d\omega \wedge \cdots \wedge d\omega)(X_1, X_2, \ldots, X_{2k+1}) \\ &= 0\end{aligned}$$

since $\iota_{X_1} \omega = 0$ and $\iota_{X_1} d\omega = 0$. This is obviously a contradiction.

We have shown that S defines an involutive distribution of dimension $n - p$. Consider the maximal integral submanifolds defined by the distribution. They give a fibring of a suitable neighborhood U^n of o whose fibres are of dimension $n - p$. We denote this fibring of U^n by $\pi\colon U^n \to U^p$. From the definition of S it is clear that ω can be projected onto U^p, that is, there exists a (unique) 1-form $\tilde{\omega}$ on U^p such that $\pi^*(\tilde{\omega}) = \omega$. This completes the proof of Lemma 1.

Lemma 2. *If ω is of rank n, then there exists a function f such that $df \neq 0$ and $df \wedge \omega^{(n-1)} = 0$ in a neighborhood of* o.

Proof of Lemma 2. We write

$$\omega^{(n-1)} = \sum_i (-1)^i a_i \, dx^1 \wedge \cdots \wedge \widehat{dx^i} \wedge \cdots \wedge dx^n.$$

Then the equation $df \wedge \omega^{(n-1)} = 0$ reduces to

$$\sum_i a_i \frac{\partial f}{\partial x^i} = 0.$$

This partial differential equation admits (infinitely many) solutions f such that $df \neq 0$ in a neighborhood of o.

Lemma 3. *If ω is of rank p, then there exists a function f such that $df \neq 0$ and $df \wedge \omega^{(p-1)} = 0$.*

Proof of Lemma 3. This is immediate from Lemmas 1 and 2.

It is also evident that in Lemma 3 we can choose f in such a way that it is constant on each fibre in the fibering $\pi: U^n \to U^p$ given in Lemma 1.

We say in general that a form ω on M depends on parameters $t_1, \ldots, t_m$ if it is a form on $M \times N$, where N is the space of parameters $t_1, \ldots, t_m$, such that its restriction to $\{p\} \times N$ vanishes for every point p of M, i.e., such that it does not involve $dt_1, \ldots, dt_m$ when expressed in a local coordinate system. The exterior derivative $d\omega$ of such a form ω on M which depends on the parameters $t_1, \ldots, t_m$ is taken considering $t_1, \ldots, t_m$ as constants. In other words, we take the ordinary exterior derivative of ω as a form on $M \times N$ and then delete the terms containing $dt_1, \ldots, dt_m$. Looking at the proofs of Lemmas 1 and 2, it is clear that Lemma 3 may be generalized to the case where ω depends on parameters. We have

Lemma 4. *If, in Lemma 3, ω depends on m parameters $t_1, \ldots, t_m$, then we can find a function f which depends differentiably on $t_1, \ldots, t_m$.*

Making use of these lemmas, we shall now prove the theorem. We denote by y^1 the function f obtained in Lemma 3 and try to find a new function f such that

$$df \wedge dy^1 \neq 0 \quad \text{and} \quad df \wedge dy^1 \wedge \omega^{(p-3)} = 0.$$

Choose a local coordinate system $y^1, u^2, \ldots, u^n$ around o; since $dy^1 \neq 0$, such a local coordinate system exists. We write

$$\omega = h\, dy^1 + \varphi,$$

where φ does not contain dy^1. We consider a neighborhood of o of the form $U_1 \times U_2$ such that y^1 is a coordinate system in U_1 and $u^2, \ldots, u^n$ is a coordinate system in U_2 in a natural manner. Then φ may be considered as a 1-form on U_2 which depends on the parameter y^1. We shall denote φ by $\tilde{\varphi}$ when we consider it as a form U_2 which depends on the parameter y^1. By direct calculation we obtain

$$dy^1 \wedge \omega^{(j)} = dy^1 \wedge \tilde{\varphi}^{(j)} \qquad \text{for every } j \geqq 1.$$

Setting $j=p-1$, we see that $\tilde{\varphi}^{(p-1)}=0$. By direct calculation we obtain also

$$\omega^{(p)}=(k\alpha\wedge dy^1+d\tilde{\varphi})\wedge\tilde{\varphi}^{(p-2)},$$

where $p=2k$ or $p=2k+1$ according as p is even or odd and α is a 1-form such that $d\omega=\alpha\wedge dy^1+d\tilde{\varphi}$. Since $\omega^{(p)}\neq 0$, we have $\tilde{\varphi}^{(p-2)}\neq 0$. Hence $\tilde{\varphi}$ is of rank $p-2$. Applying Lemma 4 to $\tilde{\varphi}$, we obtain a function f on U_2 such that $df\neq 0$ and $df\wedge\tilde{\varphi}^{(p-3)}=0$. We can extend f to a function on $U_1\times U_2$ in a natural manner; the extended function will be denoted also by f. Then $dy^1\wedge df\neq 0$ and $dy^1\wedge df\wedge\omega^{(p-3)}=dy^1\wedge df\wedge\tilde{\varphi}^{(p-3)}=0$. Setting $y^2=f$, we may write

$$dy^1\wedge dy^2\neq 0 \quad\text{and}\quad dy^1\wedge dy^2\wedge\omega^{(p-3)}=0.$$

Continuing in this way we obtain functions $y^1,\dots,y^k$ such that

$$dy^1\wedge\cdots\wedge dy^k\neq 0$$

and

$$dy^1\wedge\cdots\wedge dy^k\wedge\omega^{(p-2k+1)}=0.$$

According as $p=2k$ or $p=2k+1$, we have

$$dy^1\wedge\cdots\wedge dy^k\wedge\omega=0 \quad \text{if } p=2k,$$
$$dy^1\wedge\cdots\wedge dy^k\wedge d\omega=0 \quad \text{if } p=2k+1.$$

If $p=2k$, we have therefore

$$\omega=z^1\,dy^1+z^2\,dy^2+\cdots+z^k\,dy^k,$$

where $z^1,\dots,z^k$ are functions. Since $\omega^{(2k)}=d\omega\wedge\cdots\wedge d\omega\neq 0$, it follows that $dy^1\wedge\cdots\wedge dy^k\wedge dz^1\wedge\cdots\wedge dz^k\neq 0$. Setting

$$x^1=z^1,\quad x^2=y^1,\dots,x^{2k-1}=z^k,\quad x^{2k}=y^k$$

and defining $x^{2k+1},\dots,x^n$ suitably, we obtain a local coordinate system $x^1,\dots,x^n$ with the desired property. If $p=2k+1$, we take a local coordinate system $y^1,\dots,y^k$, $u^{k+1},\dots,u^n$; since $dy^1\wedge\cdots\wedge dy^k\neq 0$, such a coordinate system exists. We then write

$$\omega=h_1\,dy^1+\cdots+h_k\,dy^k+\psi,$$

where ψ is a 1-form which does not involve $dy^1,\dots,dy^k$. As we have explained before Lemma 4, ψ may be considered as a 1-form on the space of $u^{k+1},\dots,u^n$ which depends on the parameter $y^1,\dots,y^k$ and will be denoted by $\tilde{\psi}$ when it is considered as a form in $u^{k+1},\dots,u^n$. Hence $d\tilde{\psi}$ is obtained from $d\psi$ deleting the terms involving $dy^1,\dots,dy^k$. Since $0=dy^1\wedge\cdots\wedge dy^k\wedge d\omega=dy^1\wedge\cdots\wedge dy^k\wedge d\psi$, it follows that $d\tilde{\psi}=0$. By Poincaré lemma, there exists a function g of $y^1,\dots,y^k$, $u^{k+1},\dots,u^n$ such that

$$dg = \psi + g_1\, dy^1 + \cdots + g_k\, dy^k,$$

where $g_i = \partial g / \partial y^i$. Hence we have

$$\omega = (h_1 - g_1)\, dy^1 + \cdots + (h_k - g_k)\, dy^k + dg.$$

If we set $x^1 = h_1 - g_1$, $x^2 = y^1, \ldots, x^{2k-1} = h_k - g_k$, $x^{2k} = y^k$, $x^{2k+1} = g$, then $dx^1 \wedge dx^2 \wedge \cdots \wedge dx^{2k+1} \neq 0$ since $\omega^{(2k+1)} \neq 0$. Choosing $x^{2k+2}, \ldots, x^n$ suitably, we obtain a local coordinate system with the desired property. q.e.d.

Let Ω be a 2-form defined on a manifold M of dimension n. We say that Ω is of *rank* $2p$ at a point o if $\Omega^p = \Omega \wedge \cdots \wedge \Omega$ (p times) is nonzero but $\Omega^{p+1} = 0$ at o. We say that Ω is of *maximal rank* if it is of rank n. As an application of Theorem 3.1, we prove

Theorem 2. *If a closed 2-form Ω is of rank $2p$ in a neighborhood of a point* o, *then there exists a local coordinate system $x^1, \ldots, x^n$ around* o *such that*

$$\Omega = dx^1 \wedge dx^2 + \cdots + dx^{2p-1} \wedge dx^p.$$

Proof. Since Ω is closed, by Poincaré lemma we have $\Omega = d\omega$, where ω is a 1-form defined in a neighborhood of o. Then ω is of rank either $2p$ or $2p+1$. By Theorem 3.1 there exists a local coordinate system $x^1, \ldots, x^n$ around o such that either

$$\omega = x^1\, dx^2 + \cdots + x^{2p-1}\, dx^{2p}$$

or

$$\omega = x^1\, dx^2 + \cdots + x^{2p-1}\, dx^{2p} + dx^{2p+1}.$$

Then it is evident that Ω has a desired expression. q.e.d.

Theorem 3.1 is due to Frobenius and Darboux. We followed the presentation of E. Cartan in [9] to which we refer the reader for relevant references. See also Arens [2].

2. Some Integral Formulas

We prove first the following integral formula of Yano [6] (see also Yano and Bochner [1]).

Theorem 1. *Let M be a compact, orientable Riemannian manifold with Riemannian connection ∇ and Ricci tensor S. Then, for every vector field X on M, we have*

$$\int_M (S(X,X) + \operatorname{trace}(A_X A_X) - (\operatorname{div} X)^2)\, dv = 0,$$

where A_X *is the field of linear endomorphisms defined by* $A_X Y = -\nabla_Y X$, $\operatorname{div} X$ *is the divergence of* X *and* dv *denotes the volume element of* M.

Proof. The proof is in terms of local coordinates. It suffices to express the integrand as the divergence of a vector field. Let ξ^i be the components of X with respect to a local coordinate system $x^1, \ldots, x^n$. Then

$$\begin{aligned}\operatorname{div}(A_X X) &= -\sum_{i,j} \nabla_i(\nabla_j \xi^i \cdot \xi^j)\\ &= -\sum_{i,j} (\nabla_i \nabla_j \xi^i \cdot \xi^j + \nabla_j \xi^i \cdot \nabla_i \xi^j)\\ &= -\sum_{i,j} (\nabla_j \nabla_i \xi^i \cdot \xi^j + R^i_{kij} \xi^k \xi^j + \nabla_j \xi^i \cdot \nabla_i \xi^j)\\ &= -\sum_{i,j} (\nabla_j \nabla_i \xi^i \cdot \xi^j + R_{kj} \xi^k \xi^j + \nabla_j \xi^i \cdot \nabla_i \xi^j)\end{aligned}$$

and

$$\begin{aligned}\operatorname{div}((\operatorname{div} X) X) &= \sum_j \nabla_j (\sum_j \nabla_i \xi^i \cdot \xi^j)\\ &= \sum_{i,j} (\nabla_j \nabla_i \xi^i \cdot \xi^j + \nabla_i \xi^i \cdot \nabla_j \xi^j).\end{aligned}$$

From these two equalities, it is clear that the integrand in Theorem 1 is equal to $-\operatorname{div}(A_X X) - \operatorname{div}((\operatorname{div} X) X)$. q.e.d.

In terms of a local coordinate system, the formula in Theorem 1 reads as follows:

$$\int_M \sum_{i,j} (R_{ij} \xi^i \xi^j + \nabla_j \xi^i \cdot \nabla_i \xi^j - \nabla_j \xi^j \cdot \nabla_i \xi^i) = 0.$$

From Theorem 1, we obtain immediately the following three formulas:

Corollary. *With the same notations as in Theorem* 1, *we have*

(1) $$\int_M \{S(X,X) + \operatorname{trace}(A_X \cdot {}^t A_X) + \tfrac{1}{2} \operatorname{trace}((A_X - {}^t A_X)^2) - (\operatorname{div} X)^2\}\, dv = 0;$$

(2) $$\int_M \{S(X,X) - \operatorname{trace}(A_X \cdot {}^t A_X) - \tfrac{1}{2} \operatorname{trace}((A_X + {}^t A_X)^2) - (\operatorname{div} X)^2\}\, dv = 0;$$

(3) $$\int_M \Big\{ S(X,X) - \operatorname{trace}(A_X \cdot {}^t A_X) + \tfrac{1}{2} \operatorname{trace}\Big(\Big(A_X + {}^t A_X - \frac{2}{n} (\operatorname{div} X) I_n\Big)^2\Big) - \frac{n-2}{n} (\operatorname{div} X)^2 \Big\}\, dv = 0.$$

We shall now prove a Kählerian analog of Theorem 1.

Theorem 2. *Let* M *be a compact Kähler manifold with Ricci tensor* S. *For any complex vector fields* X *and* Y *of type* (1, 0), *we have*

$$\int_M \{S(X, \overline{Y}) + \operatorname{trace}(A''_X \cdot \overline{A''_Y}) - (\operatorname{div} X)(\operatorname{div} \overline{Y})\}\, dv = 0,$$

where A''_X is the field of linear transformations sending a vector field $\bar{Z}$ of type (0, 1) *into the vector field* $-\nabla_{\bar{Z}} X$ *of type* (1, 0), *i.e.*,

$$A''_X(\bar{Z}) = -\nabla_{\bar{Z}} X.$$

Before we start the proof, we remark that $\overline{A''_Y}$ sends a vector field of type (1, 0) into a vector field of type (0, 1) so that $A''_X \cdot \overline{A''_Y}$ sends a vector field of type (1, 0) into a vector field of the same type. We note also that, by setting $X = Y$ in the formula above, we obtain a Kählerian analog of the formula in Theorem 1.

Proof. In terms of a local coordinate system $z^1, \ldots, z^n$, let

$$X = \sum \xi^\alpha \frac{\partial}{\partial z^\alpha} \quad \text{and} \quad Y = \sum \eta^\alpha \frac{\partial}{\partial z^\alpha}.$$

Then the components of A''_X are given by $-\nabla_{\bar{\beta}} \xi^\alpha$. The components of the vector field $A''_X(\bar{Y}) = -\nabla_{\bar{Y}} X$ of type (1, 0) are given by $-\sum \nabla_{\bar{\beta}} \xi^\alpha \overline{\eta^\beta}$. A calculation similar to the one in the proof of Theorem 1 yields

$$-\sum \nabla_\alpha(\nabla_{\bar{\beta}} \xi^\alpha \cdot \overline{\eta^\beta}) = -\sum (\nabla_{\bar{\beta}} \nabla_\alpha \xi^\alpha \cdot \overline{\eta^\beta} + R_{\alpha\bar{\beta}} \xi^\alpha \overline{\eta^\beta} + \nabla_{\bar{\beta}} \xi^\alpha \cdot \nabla_\alpha \overline{\eta^\beta})$$

and

$$\sum \nabla_{\bar{\beta}}(\nabla_\alpha \xi^\alpha \cdot \overline{\eta^\beta}) = \sum (\nabla_{\bar{\beta}} \nabla_\alpha \xi^\alpha \cdot \overline{\eta^\beta} + \nabla_\alpha \xi^\alpha \cdot \nabla_{\bar{\beta}} \overline{\eta^\beta}).$$

By adding these two equalities and integrating the resulting equality over M, we obtain

$$\int_M \sum (R_{\alpha\bar{\beta}} \xi^\alpha \overline{\eta^\beta} + \nabla_{\bar{\beta}} \xi^\alpha \cdot \nabla_\alpha \overline{\eta^\beta} - \nabla_\alpha \xi^\alpha \cdot \nabla_{\bar{\beta}} \overline{\eta^\beta})\, dv = 0,$$

which is precisely the formula in Theorem 2. q.e.d.

Let X be a vector field on a Riemannian manifold M. Let ξ be the 1-form corresponding to X under the duality defined by the Riemannian metric. If we denote by Δ the Laplacian, then $\Delta\xi$ is a 1-form. We denote by ΔX the vector field corresponding to $\Delta\xi$. We state

Theorem 3. *Let M be a compact orientable Riemannian manifold with Ricci tensor S. Then, for any vector fields X and Y on M, we have*

$$\int_M \{-(\Delta X, Y) + S(X, Y) + (\nabla X, \nabla Y)\}\, dv = 0.$$

Proof. Let ξ^i and η^i be the components of X and Y with respect to a local coordinate system $x^1, \ldots, x^n$. Combining

$$\sum \nabla_i(\nabla^i \xi^j \cdot \eta_j) = \sum (\nabla_i \nabla^i \xi^j \cdot \eta_j + \nabla^i \xi^j \cdot \nabla_i \eta_j)$$

with the formula for the Laplacian (see Appendix 3), we obtain

$$\sum \nabla_i(\nabla^i \xi^j \cdot \eta_j) = -\sum (\Delta\xi)^j \cdot \eta_j + \sum R_{ij} \xi^i \eta^j + \sum \nabla^i \xi^j \cdot \nabla_i \eta_j.$$

This, integrated over M, yields the desired formula. q.e.d.

Let X be a complex vector field of type (1, 0) on a Kähler manifold M and let ξ be the corresponding (0, 1)-form. (If $X=\sum \xi^\alpha \frac{\partial}{\partial z^\alpha}$, then $\xi=\sum \xi_{\bar\alpha}\, d\bar z^\alpha$, where $\xi_{\bar\alpha}=\sum g_{\beta\bar\alpha}\xi^\beta$.) Let $\Delta''=\delta'' d''+d''\delta''$ be the d''-Laplacian. Then $\Delta''\xi$ is a (0, 1)-form. We denote by $\Delta'' X$ the corresponding vector field of type (1, 0). As a Kählerian analog of Theorem 3, we obtain

Theorem 4. *Let M be a compact Kähler manifold with Ricci tensor S. For any complex vector fields X and Y of type* (1, 0), *we have*

$$\int_M \{-(\Delta'' X, \bar Y)+S(X, \bar Y)+(\nabla'' X, \overline{\nabla'' Y})\}\, dv=0.$$

Proof. From

$$\sum \nabla_\alpha(\nabla^\alpha \xi^\beta\cdot \overline{\eta_\beta})=\sum(\nabla_\alpha\nabla^\alpha\xi^\beta\cdot\overline{\eta_\beta}+\nabla^\alpha\xi^\beta\cdot\nabla_\alpha\overline{\eta_\beta})$$

and from the formula for the d''-Laplacian (see Appendix 3), we obtain the desired formula in the same way as in the proof of Theorem 3. q.e.d.

Expressed in terms of local coordinates, we obtain

$$\int_M \{-\sum(\Delta''\xi)_{\bar\beta}\bar\eta^\beta+\sum R_{\alpha\beta}\xi^\alpha\bar\eta^\beta+\sum\nabla_{\bar\beta}\xi^\alpha\cdot\overline{\nabla^\beta\xi_{\bar\alpha}}\}\, dv=0.$$

3. Laplacians in Local Coordinates

Let ω be a p-form on a Riemannian manifold M. Let

$$\omega=\frac{1}{p!}\sum\omega_{i_1 i_2\dots i_p}\, dx^{i_1}\wedge dx^{i_2}\wedge\dots\wedge dx^{i_p}$$

be a local expression for ω. Then

$$d\omega=\frac{1}{(p+1)!}\sum(\nabla_i\omega_{i_1 i_2\dots i_p}-\nabla_{i_1}\omega_{i i_2\dots i_p}-\dots-\nabla_{i_p}\omega_{i_1 i_2\dots i_{p-1} i})$$
$$\cdot\, dx^i\wedge dx^{i_1}\wedge\dots\wedge dx^{i_p},$$

and

$$\delta\omega=-\frac{1}{(p-1)!}\sum\nabla_i\omega^i_{i_2\dots i_p}\, dx^{i_2}\wedge\dots\wedge dx^{i_p}.$$

In particular, let ω be a 1-form so that

$$\omega=\sum\omega_i\, dx^i.$$

Then

$$d\omega=\tfrac{1}{2}\sum(\nabla_i\omega_j-\nabla_j\omega_i)\, dx^i\wedge dx^j,$$
$$\delta d\omega=-\sum\nabla_i(\nabla^i\omega_j-\nabla_j\omega^i)\, dx^j,$$
$$\delta\omega=-\sum\nabla_i\omega^i,$$
$$d\delta\omega=-\sum\nabla_j\nabla_i\omega^i\, dx^j.$$

Hence,

$$\begin{aligned}\Delta\omega &= \delta d\omega + d\delta\omega \\ &= \sum(-\nabla_i \nabla^i \omega_j + \nabla_i \nabla_j \omega^i - \nabla_j \nabla_i \omega^i)\, dx^j \\ &= \sum(-\nabla_i \nabla^i \omega_j + R_{ij}\omega^i)\, dx^j.\end{aligned}$$

For a similar expression for a p-form ω, see Yano [9; p. 67].

Let ω be a (p, q)-form on a Kähler manifold M. Let

$$\omega = \frac{1}{p!\,q!} \sum \omega_{\alpha_1 \dots \alpha_p \bar{\beta}_1 \dots \bar{\beta}_q}\, dz^{\alpha_1} \wedge \dots \wedge dz^{\alpha_p} \wedge d\bar{z}^{\beta_1} \wedge \dots \wedge d\bar{z}^{\beta_q}$$

be a local expression for ω. Then

$$\begin{aligned}d''\omega = \frac{1}{p!\,(q+1)!} &\sum \Big(\nabla_{\bar{\beta}}\, \omega_{A \bar{\beta}_1 \dots \bar{\beta}_q} - \sum_{k=1}^{q} \nabla_{\bar{\beta}_k}\, \omega_{A \bar{\beta}_1 \dots \bar{\beta}_{k-1} \bar{\beta} \bar{\beta}_{k+1} \dots \bar{\beta}_q}\Big) \\ &\cdot dz^A \wedge d\bar{z}^{\beta} \wedge d\bar{z}^{\beta_1} \wedge \dots \wedge d\bar{z}^{\beta_q},\end{aligned}$$

where A stands for $\alpha_1 \dots \alpha_p$ and dz^A for $dz^{\alpha_1} \wedge \dots \wedge dz^{\alpha_q}$. We have

$$\delta''\omega = -\frac{1}{p!\,(q-1)!} \sum \nabla_{\beta}\, \omega_{A}{}^{\beta}{}_{\bar{\beta}_2 \dots \bar{\beta}_q}\, dz^A \wedge d\bar{z}^{\beta_2} \wedge \dots \wedge d\bar{z}^{\beta_q}.$$

Hence, if ω is a $(0, 1)$-form and

$$\omega = \sum \omega_{\bar{\alpha}}\, d\bar{z}^{\alpha},$$

then

$$\begin{aligned}d''\omega &= \tfrac{1}{2}\sum (\nabla_{\bar{\alpha}}\omega_{\bar{\beta}} - \nabla_{\bar{\beta}}\omega_{\bar{\alpha}})\, d\bar{z}^{\alpha} \wedge dz^{\beta}, \\ \delta'' d''\omega &= -\sum \nabla_{\alpha}(\nabla^{\alpha}\omega_{\bar{\beta}} - \nabla_{\bar{\beta}}\omega^{\alpha})\, d\bar{z}^{\beta} \\ \delta''\omega &= -\sum \nabla_{\alpha}\omega^{\alpha}, \\ d''\delta''\omega &= -\sum \nabla_{\bar{\beta}}\nabla_{\alpha}\omega^{\alpha}\, d\bar{z}^{\beta}.\end{aligned}$$

Hence,

$$\begin{aligned}\Delta''\omega &= (\delta'' d'' + d''\delta'')\,\omega \\ &= \sum(-\nabla_{\alpha}\nabla^{\alpha}\omega_{\bar{\beta}} + \nabla_{\alpha}\nabla_{\bar{\beta}}\omega^{\alpha} - \nabla_{\bar{\beta}}\nabla_{\alpha}\omega^{\alpha})\, d\bar{z}^{\beta} \\ &= \sum(-\nabla_{\alpha}\nabla^{\alpha}\omega_{\bar{\beta}}\, d\bar{z}^{\beta} + R_{\alpha\bar{\beta}}\,\omega^{\alpha}\, d\bar{z}^{\beta}).\end{aligned}$$

It is known that, for a Kähler manifold,

$$\Delta = 2\Delta''.$$

It is therefore possible to derive the fomula above for $\Delta''\omega$ from the formula for $\Delta\omega$.

4. A Remark on $d' d''$-Cohomology

We prove the following well known

Theorem 1. *Let θ and θ' be closed real (p, p)-forms on a compact Kähler manfiold M. Then $\theta \sim \theta'$ (cohomologous to each other) if and only if there exists a real $(p-1, p-1)$-form φ on M such that*

$$\theta - \theta' = i\, d' d'' \varphi .$$

Proof. If $\theta - \theta' = i\, d' d'' \varphi$, then $\theta - \theta' = d(i\, d'' \varphi)$ and hence $\theta \sim \theta'$. (This implication is purely local and is valid for any complex manifold.)

To prove the converse, let $\eta = \theta - \theta'$ and assume $\eta \sim 0$ so that $\eta = d\alpha$, where α is a real $(2p-1)$-form. Let $\alpha = \bar{\beta} + \beta$, where β is a $(p-1, p)$-form and $\bar{\beta}$ is its complex conjugate. Then

$$\eta = d\alpha = d' \bar{\beta} + (d'' \bar{\beta} + d' \beta) + d'' \beta ,$$

where $d' \bar{\beta}$ is of bidegree $(p+1, p-1)$, $(d'' \bar{\beta} + d' \beta)$ of bidegree (p, p), and $d'' \beta$ of bidegree $(p-1, p+1)$. Hence,

$$\eta = d\alpha = d'' \bar{\beta} + d' \beta, \qquad d' \bar{\beta} = 0, \qquad d'' \beta = 0 .$$

We may write

$$\beta = H\beta + d'' \gamma ,$$

where γ is a $(p-1, p-1)$-form. Then

$$\bar{\beta} = \overline{H\beta} + d' \bar{\gamma} .$$

Hence,

$$\begin{aligned} \eta &= d'' \bar{\beta} + d' \beta = d'' d' \bar{\gamma} + d' d'' \gamma = d' d'' (\gamma - \bar{\gamma}) \\ &= i\, d' d'' \varphi , \end{aligned}$$

where $\varphi = -i(\beta - \bar{\beta})$ is a real $(p-1, p-1)$-form. q.e.d.

Bibliography

Accola, R. D. M.

[1] Automorphisms of Riemann surfaces. J. Analyse Math. **18**, 1–5 (1967) (MR 35 #4398).

[2] On the number of automorphisms of a closed Riemann surfaces. Trans. Amer. Math. Soc. **131**, 398–408 (1968) (MR 36 #5333).

Aeppli, A.

[1] Some differential geometric remarks on complex homogeneous manifolds. Arch. Math. (Basel) **16**, 60–68 (1965) (MR 32 #464).

Ambrose, W., Singer, I. M.

[1] On homogeneous Riemannian manifolds. Duke Math. J. **25**, 647–669 (1958) (MR 21 #1628).

Andreotti, A.

[1] Sopra le superficie algebriche che posseggono transformazioni birazionali in se, Univ. Roma Ist. Naz. Alta Mat. Rend. Mat e Appl. **9**, 255–279 (1950) (MR 14, 20).

Andreotti, A., Salmon, P.

[1] Anelli con unica decomposabilità in fattori primi ed un problema di intersezione complete. Monatsh. Math. **61**, 97–142 (1957) (MR 21 #3415).

Apte, M., Lichnerowicz, A.

[1] Sur les transformations affines d'une variété presque hermitienne compacte. C. R. Acad. Sci. Paris Sér. A-B **242**, 337–339 (1956) (MR 17, 787).

Arens, R.

[1] Topologies for hemeomorphism groups. Ann. of Math. **68**, 593–610 (1946) (MR 8, 479).

[2] Normal forms for a Pfaffian. Pacific J. Math. **14**, 1–8 (1964) (MR 30 #2424).

Atiyah, M., Bott, R.

[1] A Lefschetz fixed point formula for elliptic complexes. I. Ann. of Math. **86**, 374–407 (1967) (MR 35 #3701). II. Ann. of Math. **88**, 451–491 (1968) (MR 38 #731).

Atiyah, M., Hirzebruch, F.

[1] Spin manifolds and group actions.

Atiyah, M., Singer, I. M.

[1] The index of elliptic operators. III. Ann. of Math. **87**, 536–604 (1968) (MR 38 #5245).

Aubin, T.

[1] Métriques riemanniennes et courbure. J. Differential Geometry **4**, 383–424 (1970).

Ba, B.

[1] Structures presque complexes, structures conformes et dérivations. Cahiers Topologie Géom. Différentielle **8**, 1–74 (1966) (MR 37 #5809).

Barbance, C.

[1] Transformations conformes d'une variété riemannienne compacte. C. R. Acad. Sci. Paris Sér. A-B **266**, A149–152 (1968) (MR 38 #5137).

Barros, C. de

[1] Sur la géométrie différentielle des formes différentielles extérieures quadratiques. Atti Convergno Internaz. Geom. Diff. (Bologna), 1967, 1–26.

Baum, P. F., Bott, R.

[1] On the zeroes of meromorphic vector fields. Essays on topology and related topics. Mémoires dédiés à G. de Rham, 1970, 29–47 (MR 41 #6248).

Baum, P. F., Cheeger, J.

[1] Infinitesimal isometries and Pontrjagin numbers. Topology **8**, 173–193 (1969) (MR 38 #6627).

Berger, M.

[1] Trois remarques sur les variétés riemanniennes à courbure positive. C. R. Acad. Sci. Paris Sér. A-B **263**, 76–78 (1966) (MR 33 #7966).

[2] Sur les variétés d'Einstein. Comptes Rendus de la IIIe Réunion du Groupement des Mathématiciens d'Expression Latine, Namur (1965), 33–55 (MR 38 #6502).

Bergmann, S.

[1] Über die Kernfunktion eines Bereiches, ihr Verhalten am Rande. J. Reine Angew. Math. **169**, 1–42 (1933) and **172**, 89–128 (1935).

Bernard, D.

[1] Sur la géométrie différentielle des *G*-structures. Ann. Inst. Fourier (Grenoble) **10**, 151–270 (1960) (MR 23 #A4094).

Bishop, R. L., Crittenden, R. J.

[1] Geometry of manifolds. New York: Academic Press 1964 (MR 29 #6401).

Bishop, R. L., O'Neill, B.

[1] Manifolds of negative curvature. Trans. Amer. Math. Soc. **145**, 1–49 (1969) (MR 40 #489).

Blanchard, A.

[1] Sur les variétés analytiques complexes. Ann. Sci. École Norm. Sup. **63**, 157–202 (1958) (MR 19, 316).

Bochner, S.

[1] Vector fields and Ricci curvature. Bull. Amer. Math. Soc. **52**, 776–797 (1946) (MR 8, 230).

[2] On compact complex manifolds. J. Indian Math. Soc. **11**, 1–21 (1947) (MR 9, 423).

[3] Tensor fields with finite bases. Ann. of Math. **53**, 400–411 (1951) (MR 12, 750).

Bochner, S., Martin, W. T.

[1] Several complex variables. Princeton Univ. Press 1948 (MR 10, 366).

Bochner, S., Montgomery, D.

[1] Locally compact groups of differentiable transformations. Ann. of Math. **47**, 639–653 (1946) (MR 8, 253).

[2] Groups on analytic manifolds. Ann. of Math. **48**, 659–669 (1947) (MR 9, 174).

[3] Groups of differentiable and real or complex analytic transformations. Ann. of Math. **46**, 685–694 (1945) (MR 7, 241).

Boothby, W. M.
[1] Homogeneous complex contact manifolds. Proc. Symp. Pure Math. vol. III, 144–154, Amer. Math. Soc. 1961 (MR 23 #A2173).
[2] A note on homogeneous complex contact manifolds. Proc. Amer. Math. Soc. **13**, 276–280 (1962) (MR 25 #582).
[3] Transitivity of automorphisms of certain geometric structures. Trans. Amer. Math. Soc. **137**, 93–100 (1969) (MR 38 # 5254).

Boothby, W. M., Kobayashi, S., Wang, H. C.
[1] A note on mappings and automorphisms of almost complex manifolds. Ann. of Math. **77**, 329–334 (1963) (MR 26 #4284).

Boothby, W. M., Wang, H. C.
[1] On contact manifolds. Ann. of Math. **68**, 721–734 (1958) (MR 22 #3015).

Borel, A.
[1] Topology of Lie groups and characteristic classes. Bull. Amer. Math. Soc. **61**, 397–432 (1955) (MR 17, 282).
[2] Some remarks about Lie groups transitive on spheres and tori. Bull. Amer. Math. Soc. **55**, 580–587 (1949) (MR 10, 680).
[3] Seminar on transformation groups. Ann. of Math., Studies No. 46, (1960) (MR 22 #7129).
[4] Groupes linéaires algébriques. Ann. of Math. **64**, 20–82 (1956) (MR 19 #1195).
[5] Linear algebraic groups. Benjamin 1969 (MR 40 #4273).

Borel, A., Narasimhan, R.
[1] Uniqueness conditions for certain holomorphic mappings. Invent. Math. **2**, 247–255 (1967) (MR 35 #414).

Borel, A., Remmert, R.
[1] Über kompakte homogene Kählersche Mannigfaltigkeiten. Math. Ann. **145**, 429–439 (1961/62) (MR 26 #3088).

Bott, R.
[1] Vector fields and characteristic numbers. Michigan Math. J. **14**, 231–244 (1967) (MR 35 #2297).
[2] A residue formula for holomorphic vector fields. J. Differential Geometry **1**, 311–330 (1967) (MR 38 #730).
[3] Non-degenerate critical manifolds. Ann. of Math. **60**, 248–261 (1954) (MR 16, 276).

Brickell, F.
[1] Differentiable manifolds with an area measure. Canad. J. Math. **19**, 540–549 (1967) (MR 35 #2253).

Busemann, H.
[1] Similarities and differentiability. Tôhoku Math. J. **9**, 56–67 (1957) (MR 20 #2772).
[2] Spaces with finite groups of motions. J. Math. Pures Appl. **37**, 365–373 (1958) (MR 21 #913).
[3] The geometry of geodesics. Academic Press 1955 (MR 17, 779).

Buttin, C.
[1] Les dérivations des champs de tenseurs et l'invariant différentiel de Schouten. C. R. Acad. Sci. Paris Sér. A-B **269**, A87–89 (1969) (MR 40 #1913).
[2] Étude d'un cas d'isomorphisme d'une algèbre de Lie filtrée avec son algèbre graduée associée. C. R. Acad. Sci. Paris Sér. A-B **264**, A496–498 (1967) (MR 39 #2826).

Calabi, E.
[1] On Kähler manifolds with vanishing canonical class. Algebraic geometry & topology. Symposium in honor of S. Lefschetz, 78–89, Princeton, 1957 (MR 19, 62).

Carathéodory, C.

[1] Über die Abbildungen, die durch Systeme von analytischen Funktionen von mehreren Veränderlichen erzeugt werden. Math. Z. **34**, 758–792 (1932).

Cartan, E.

[1] Sur les variétés à connexion affine et la théorie de la relativité généralisée. Ann. Sci. École Norm. Sup. **40**, 325–412 (1923); **41**, 1–25 (1924), and **42**, 17–88 (1925).

[2] Les espaces à connexion conforme. Ann. Soc. Po. Math. **2**, 171–221 (1923).

[3] Sur un théorème fondamental de M. H. Weyl. J. Math. Pures Appl. **2**, 167–192 (1923).

[4] Sur les variétés à connexion projective. Bull. Soc. Math. France **52**, 205–241 (1924).

[5] Sur la structure des groupes infinis de transformations. Ann. Sci. École Norm. Sup. **21**, 153–206 (1904), and **22**, 219–308 (1905).

[6] Les sous-groupes des groupes continus de transformations. Ann. Sci. École Norm. Sup. **25**, 57–194 (1908).

[7] Les groupes de transformations continus, infinis, simples. Ann. Sci. École Norm. Sup. **26**, 93–161 (1909).

[8] Leçons sur la Géométrie des Espaces de Riemann. Paris: Gauthier-Villars 1928; 2nd ed. 1946.

[9] Leçons sur les invariants intégraux. Paris: Hermann 1922.

Cartan, H.

[1] Sur les groupes de transformations analytiques. Actualités Sci. Indust. No. 198. Paris: Hermann 1935.

[2] Sur les fonctions de plusieurs variables complexes. L'itération des transformations intérieures d'un domaine borné, Math. Z. **35**, 760–773 (1932).

Chern, S.S.

[1] Pseudo-groupes continus infinis. Colloque de Géométrie Diff. Strasbourg (1953), 119–136 (MR 16, 112).

[2] The geometry of G-structures. Bull. Amer. Math. Soc. **72**, 167–219 (1966) (MR 33 #661).

Chevalier, H.

[1] Noyau sur une variété analytique. C. R. Acad. Sci. Paris Sér. A-B **262**, A 948–950 (1966) (MR 34 #3591).

Chevalley, C.

[1] Theory of Lie groups. Princeton Univ. Press 1946 (MR 7, 412).

Chu, H., Kobayashi, S.

[1] The automorphism group of a geometric structure. Trans. Amer. Math. Soc. **113**, 141–150 (1963) (MR 29 #1596).

Conner, P.E.

[1] On the action of the circle group. Michigan Math. J. **4**, 241–247 (1957) (MR 20 #3230).

Couty, R.

[1] Sur les transformations de variétés riemanniennes et kählériennes. Ann. Inst. Fourier (Grenoble) **9**, 147–248 (1959) (MR 22 #12488).

[2] Transformations projective sur un espace d'Einstein complet. C. R. Acad. Sci. Paris Sér. A-B **252**, 109–1097 (1961) (MR 28 #569).

[3] Formes projectives fermées sur un espace d'Einstein, C. R. Acad. Sci. Paris Sér. A-B **270**, A 1249–1251 (1970) (MR 42 #1008).

Dantzig, D. van, Waerden, B. L. van der

[1] Über metrisch homogene Räume. Abh. Math. Sem. Univ. Hamburg **6**, 374–376 (1928).

Dinghas, A.

[1] Verzerrungssätze bei holomorphen Abbildungen von Hauptbereichen automorpher Gruppen mehrerer komplexer Veränderlicher in einer Kähler-Mannigfaltigkeit. S.-B. Heidelberger Akad. Wiss. Math.-Natur. Kl. **1968**, 1–21 (MR 38 #6109).

Douglis, A., Nirenberg, L.

[1] Interior estimates for elliptic systems of partial differential equations. Comm. Pure Appl. Math. **8**, 503–583 (1955) (MR 17, 743).

Ebin, D. G., Marsden, J. E.

[1] Groups of diffeomorphisms and the motion of an incompressible fluid. Ann. of Math. **92**, 102–163 (1970) (MR 42 #6865).

Eckmann, B.

[1] Structures complexes et transformations infinitesimales. Convergno Intern. Geometria Differenziale, Italia, 1953, 176–184 (MR 16, 518).

Eells, J., Jr.

[1] A setting for global analysis. Bull. Amer. Math. Soc. **72**, 751–807 (1966) (MR 34 #3590).

Eells, J., Jr., Sampson, H.

[1] Harmonic mappings of Riemannian manifolds. Amer. J. Math. **86**, 109–160 (1964) (MR 29 #1603).

[2] Énergie et déformations en géométrie différentielle. Ann. Inst. Fourier (Grenoble) **14**, 61–69 (1964) (MR 30 #2529).

Egorov, I. P.

[1] On the order of the group of motions of spaces with affine connection. Dokl. Akad. Nauk SSSR **57**, 867–870 (1947) (MR 9, 468).

[2] On collineations in spaces with projective connection. Dokl. Akad. Nauk SSSR **61**, 605–608 (1948) (MR 10, 211).

[3] On the groups of motions of spaces with asymmetric affine connection. Dokl. Akad. Nauk SSSR **64**, 621–624 (1949) (MR 10, 739).

[4] On a strengthening of Fubini's theorem on the order of the group of motions of a Riemannian space. Dokl. Akad. Nauk SSSR **66**, 793–796 (1949) (MR 11, 211).

[5] On groups of motions of spaces with general asymmetrical affine connection. Dokl. Akad. Nauk SSSR **73**, 265–267 (1950) (MR 12, 636).

[6] Collineations of projectively connected spaces. Dokl. Akad. Nauk SSSR **80**, 709–712 (1951) (MR 14, 208).

[7] A tensor characterization of A_n of nonzero curvature with maximum mobility. Dokl. Akad. Nauk SSSR **84**, 209–212 (1952) (MR 14, 318).

[8] Maximally mobile L_n with a semi-symmetric connection. Dokl. Akad. Nauk SSSR **84**, 433–435 (1952) (MR 14, 318).

[9] Motions in spaces with affine connection. Dokl. Akad. Nauk SSSR **87**, 693–696 (1952) (MR 16, 171).

[10] Maximally mobile Riemannian spaces V_4 of non constant curvature. Dokl. Akad. Nauk SSSR **103**, 213–232 (1955) (MR 17, 405).

[11] Motions in affinely connected spaces. Dokl. Akad. Nauk SSSR **87**, 693–696 (1952) (MR 16, 170).

[12] Maximally mobile Einstein spaces with non constant curvature. Dokl. Akad. Nauk SSSR **145**, 975–978 (1962) (MR 25 #4472).

[13] Equi-affine spaces of third lacunary. Dokl. Akad. Nauk SSSR **108**, 1007–1010 (1956) (MR 18, 506).

[14] Riemann spaces of second lacunary. Dokl. Akad. Nauk SSSR **111**, 276–279 (1956), (MR 19, 169).

[15] Riemannian spaces of the first three lacunary types in the geometric sense. Dokl. Akad. Nauk SSSR **150**, 730–732 (1963) (MR 27 #5201).

Ehresmann, C.

[1] Les connexions infinitésimales dans un espace fibré différentiable. Colloque de Topologie, Bruxelles 1950, 29–55 (MR 13, 159).

[2] Sur les variétés presque complexes. Proc. Intern. Congr. Math. Cambridge, Mass. 1950, vol. 2, 412–419 (MR 13, 547).

[3] Introduction à la théorie des structures infinitésimales et des pseudo-groupes de Lie. Colloque de Géométrie Diff., Strasbourg, 1953, 97–110 (MR 16, 75).

[4] Sur les pseudo-groupes de Lie de type fini. C. R. Acad. Sci. Paris Sér. A-B **246**, 360–362 (1958) (MR 21 #97).

Eisenhart, L. P.

[1] Riemannian geometry. Princeton: Princeton Univ. Press 1949 (MR 11, 687).

[2] Non-Riemannian Geometry. Amer. Soc. Colloq. Publ. **8**, 1927.

[3] Continuous groups of transformations. Princeton: Princeton Univ. Press. 1933.

Floyd, E. E.

[1] On periodic maps and the Euler characteristics of associated spaces. Trans. Amer. Math. Soc. **72**, 138–147 (1952) (MR 13, 673).

Frankel, T. T.

[1] Fixed points and torsions on Kähler manifolds. Ann. of Math. **70**, 1–8 (1959) (MR 24 #A 1730).

[2] Manifolds with positive curvature. Pacific J. Math. **11**, 165–174 (1961) (MR 23 #A 600).

[3] On theorem of Hurwitz and Bochner. J. Math. Mech. **15**, 373–377 (1966) (MR 33 #675).

Freifield, C.

[1] A conjecture concerning transitive subalgebras of Lie algebras. Bull. Amer. Math. Soc. **76**, 331–333 (1970) (MR 40 #5794).

Frölicher, A.

[1] Zur Differentialgeometrie der komplexen Strukturen. Math. Ann. **129**, 50–95 (1955) (MR 16, 857).

Fubini, G.

[1] Sugli spazi che amettono un gruppo continuo di movimenti. Annali di Mat. **8**, 39–81 (1903).

Fujimoto, A.

[1] On the structure tensor of *G*-structures. Mem. Coll. Sci. Univ. Kyoto Ser. **A 18**, 157–169 (1960) (MR 22 #8522).

[2] On automorphisms of *G*-structures. J. Math. Kyoto Univ. **1**, 1–20 (1961) (MR 25 #4459).

[3] Theory of *G*-structures. A report in differential geometry [in Japanese], vol. 1, 1966.

Fujimoto, H.

[1] On the holomorphic automorphism groups of complex spaces. Nagoya Math. J. **33**, 85–106 (1968) (MR 39 #3038).

[2] On the automorphism group of a holomorphic fibre bundle over a complex space. Nagoya Math. J. **37**, 91–106 (1970) (MR 41 #3809).

[3] On holomorphic maps into a real Lie group of holomorphic transformations. Nagoya Math. J. **40**, 139–146 (1970) (MR 42 #3310).

Godbillon, C.
[1] Géométrie différentielle et mécanique analytique. Paris: Hermann 1969 (MR 39 #3416).

Goldberg, S.I.
[1] Curvature and Homology. New York: Academic Press 1962 (MR 25 #2537).

Goldberg, S.I., Kobayashi, S.
[1] The conformal transformation group of a compact Riemannian manifold. Amer. J. Math. **84**, 170–174 (1962) (MR 25 #3481).

Gottschling, E.
[1] Invarianten endlicher Gruppen und biholomorphe Abbildungen. Invent. Math. **6**, 315–326 (1969) (MR 39 #5834).
[2] Reflections in bounded symmetric domains. Comm. Pure Appl. Math. **22**, 693–714 (1969) (MR 40 #4479).

Grauert, H., Remmert, R.
[1] Über kompakte homogene komplexe Mannigfaltigkeiten. Arch. Math. **13**, 498–507 (1962) (MR 26 #3089).

Gray, J.W.
[1] Some global properties of contact structure. Ann. of Math. **69**, 421–450 (1959) (MR 22 #3016).

Gromoll, D., Wolf, J.A.
[1] Some relations between the metric structure and the algebraic structure of the fundamental group in manifolds of nonpositive curvature. Bull. Amer. Math. Soc. **77**, 545–552 (1971).

Guillemin, V.
[1] A Jordan-Hölder decomposition for a certain class of infinite dimensional Lie algebras. J. Diff. Geometry **2**, 313–345 (1966) (MR 41 #8481).
[2] Infinite dimensional primitive Lie algebra, J. Differential Geometry **4**, 257–282 (1970) (MR 42 #3132).
[3] Integrability problem for *G*-structures. Trans. Amer. Math. Soc. **116**, 544–560 (1965) (MR 34 #3475).

Guillemin, V., Quillen, D., Sternberg, S.
[1] The classification of the irreducible complex algebras of infinite type. J. Analyse Math. **18**, 107–112 (1967) (MR 36 #221).

Guillemin, V., Sternberg, S.
[1] An algebraic model of transitive differential geometry. Bull. Amer. Math. Soc. **70**, 16–47 (1964) (MR 30 #533).
[2] Deformation theory of pseudogroup structures. Mem. Amer. Math. Soc. no. 64 (1966) (MR 35 #2302).

Gunning, R.C.
[1] On Vitali's theorem for complex spaces with singularities. J. Math. Mech. **8** 133–141 (1959) (MR 21 #6446).

Gunning, R.C., Rossi, H.
[1] Analytic functions of several complex variables. Englewood Cliffs, N.J.: Prentice-Hall 1965 (MR 31 #4927).

Hano, J.-I.
[1] On affine transformations of a Riemannian manifold. Nagoya Math. J. **9**, 99–109 (1955) (MR 17, 891).
[2] On compact complex coset spaces of reductive Lie groups. Proc. Amer. Math. Soc. **15**, 159–163 (1964) (MR 28 #1258).

Hano, J.-I., Kobayashi, S.
[1] A fibering of a class of homogeneous complex manifolds. Trans. Amer. Math. Soc. **94**, 233–243 (1960) (MR 22 #5990).

Hano, J.-I., Matsushima, Y.
[1] Some studies of Kählerian homogeneous spaces. Nogoya Math. J. **11**, 77–92 (1957) (MR 18, 934).

Hano, J.-I., Morimoto, A.
[1] Note on the group of affine transformations of an affinely connected manifold. Nagoya Math. J. **8**, 71–81 (1955) (MR 16, 1053).

Hartman, P.
[1] On homotopic harmonic maps. Canad. J. Math. **19**, 673–687 (1967) (MR 35 #4856).

Hatakeyama, Y.
[1] Some notes on the group of automorphisms of contact and symplectic structures. Tôhoku Math. J. **18**, 338–347 (1966) (MR 34 #6679).

Hawley, N.S.
[1] A theorem on compact complex manifolds. Ann. of Math. **52**, 637–641 (1950) (MR 12, 603).

Helgason, S.
[1] Differential Geometry and Symmetric Spaces. New York: Academic Press 1962 (MR 26 #2986).

Hermann, R.
[1] Cartan connections and the equivalence problem for geometric structures. Contributions to Diff. Equations **3**, 199–248 (1964) (MR 29 #2741).

Hirzebruch, F.
[1] Topological Methods in Algebraic Geometry. Berlin-Heidelberg-New York: Springer 1966 (MR 34 #2573).

Hochschild, G.
[1] The structure of Lie Groups. San Francisco: Holden Day 1965 (MR 34 #7696).

Holmann, H.
[1] Komplexe Räume mit komplexen Transformationsgruppen. Math. Ann. **150**, 327–360 (1963) (MR 27 #776).

Howard, A.
[1] Holomorphic vector fields on algebraic manifolds. Amer. J. Math., to appear.

Hsiang, W.Y.
[1] The natural metric on $SO(n)/SO(n-2)$ is the most symmetric metric. Bull. Amer. Math. Soc. **73**, 55–58 (1967) (MR 35 #939).
[2] On the bound of the dimensions of the isometry groups of all possible riemannian metrics on an exotic sphere. Ann. of Math. **85**, 351–358 (1967) (MR 35 #4935).
[3] On the degree of symmetry and the structure of highly symmetric manifolds. Tamkang J. of Math., Tamkang College of Arts & Sci., Taipei, **2**, 1–22 (1971).

Hsiung, C.C.
[1] Vector fields and infinitesimal transformations on Riemannian manifolds with boundary. Bull. Soc. Math. France **92**, 411–434 (1964) (MR 31 #2693).

Huang, W.H.
[1] Lefschetz numbers for Riemannian manifolds. To appear.

Hurwitz, A.
[1] Über algebraische Gebilde mit eindeutigen Transformationen in sich. Math. Ann. **41**, 403–442 (1893).
[2] Über diejenigen algebraischen Gebilde, welche eindeutige Transformationen in sich zu lassen. Math. Ann. **32**, 290–308 (1887).

Illusie, L.
[1] Nombres de Chern et groupes finis. Topology **7**, 255–269 (1968) (MR 37 #4822).

Ishida, M.
[1] On groups of automorphisms of algebraic varieties. J. Math. Soc. Japan **14**, 276–283 (1962) (MR 26 #1315).

Ishihara, S.
[1] Homogeneous Riemannian spaces of four dimension. J. Math. Soc. Japan **7**, 345–370 (1955) (MR 18, 599).
[2] Groups of projective transformations on a projectively connected manifold. Japan. J. Math. **25**, 37–80 (1955) (MR 18, 599).
[3] Groups of projective transformations and groups of conformal transformations. J. Math. Soc. Japan **9**, 195–227 (1957) (MR 20 #311).

Ishihara, S., Obata, M.
[1] Affine transformations in a Riemannian manifold. Tôhoku Math. J. **7**, 146–150 (1955) (MR 17, 1128).

Jänich, K.
[1] Differentierbare *G*-Mannigfaltigkeiten. Lectures notes in Math. No. 59. Berlin-Heidelberg-New York: Springer 1968 (MR 37 #4835).

Jensen, G.R.
[1] Homogeneous Einstein spaces of dimension four. J. Differential Geometry **3**, 309–350 (1969) (MR 41 #6100).
[2] The scalar curvature of left-invariant Riemannian metrics. Indiana Univ. Math. J. **20**, 1125–1144 (1971).

Kac, V.G.
[1] Simple graded Lie algebras of finite growth. Izv. Akad. Nauk SSSR, Ser. Mat. **32**, 1323–1367 (1968) (MR 41 #4590).

Kaneyuki, S.
[1] On the automorphism groups of homogeneous bounded domains. J. Fac. Sci. Univ. Tokyo, Sect. I **14**, 89–130 (1967) (MR 37 #3056).

Kantor, I.L.
[1] Infinite dimensional simple graded Lie algebras. Dokl. Akad. Nauk SSSR **179**, 534–537 (1968) (MR 37 #1418). Soviet Math. Dokl. **9**, 409–412 (1968).
[2] Transitive differential groups and invariant connections in homogeneous spaces. Trudy Sem. Vektor. Tenzor. Anal. **13**, 310–398 (1966) (MR 36 #1314).

Kantor, L.L., Sirota, A.I., Solodovnikov, A.S.
[1] A class of symmetric spaces with an extensive group of motions and a generalization of the Poincaré model. Dokl. Akad. Nauk SSSR **173**, 511–514 (1967). Soviet Math. Dokl. **8**, 423–426 (1967) (MR 35 #3607).

Kaup, W.

[1] Reelle Transformationsgruppen und invariante Metriken auf komplexen Räumen. Invent. Math. **3**, 43–70 (1967) (MR 35 #6865).

[2] Holomorphe Abbildungen in hyperbolische Räume. Centro Internaz. Mat. Estivo, 1967, pp. 111–123 (MR 39 #478).

[3] Holomorphic mappings of complex spaces. Symp. Math., vol. II (INDAM, Roma, 1968), 333–340. Academic Press 1969 (MR 40 #5906).

[4] Hyperbolische komplexe Räume. Ann. Inst. Fourier (Grenoble) **18**, 303–330 (1968) (MR 39 #7138).

Kaup, W., Matsushima, Y., Ochiai, T.

[1] On the automorphisms and equivalences of generalized Siegel domains. Amer. J. Math. **92**, 475–498 (1970) (MR 42 #2029).

Kerner, H.

[1] Über die Automorphismengruppen kompakter komplexer Räume. Arch. Math. **11**, 282–288 (1960) (MR 22 #8533).

Klingenberg, W.

[1] Eine Kennzeichnung der Riemannschen sowie der Hermiteschen Mannigfaltigkeiten. Math. Z. **70**, 300–309 (1959) (MR 21 #3891).

Knebelman, M. S.

[1] Collineations of projectively related affine connections. Ann. of Math. **29**, 389–394 (1928).

[2] Collineations and motions in generalized spaces. Amer. J. Math. **51**, 527–564 (1929).

[3] On groups of motions in related spaces. Amer. J. Math. **52**, 280–282 (1930).

Kobayashi, E. T.

[1] A remark on the Nijenhuis tensor. Pacific J. Math. **12**, 963–977 and 1467 (1962) (MR 27 #678).

[2] A remark on the existence of a *G*-structure. Proc. Amer. Math. Soc. **16**, 1329–1331 (1965) (MR 35 #2235).

Kobayashi, S.

[1] Le groupe de transformations qui laissent invariant un parallélisme. Colloque de Topologie. Strasbourg: Ehresmann 1954 (MR 19, 576).

[2] Groupes de transformations qui laissent invariante une connexion infinitésimale, C. R. Acad. Sci. Paris Sér. A-B **238**, 644–645 (1954) (MR 15, 742).

[3] A theorem on the affine transformation group of a Riemannian manifold. Nagoya Math. J. **9**, 39–41 (1955) (MR 17, 892).

[4] Theory of connections. Ann. Mat. Pura Appl. **43**, 119–194 (1957) (MR 20 # 2760).

[5] Fixed points of isometries. Nagoya Math. J. **13**, 63–68 (1958) (MR 21 #2276).

[6] Geometry of bounded domains. Trans. Amer. Math. Soc. **92**, 267–290 (1959) (MR 22 #3017).

[7] On the automorphism group of a certain class of algebraic manifolds. Tôhoku Math. J. **11**, 184–190 (1959) (MR 22 #3014).

[8] Frame bundles of higher order contact. Proc. Symp. Pure Math. vol. 3, Amer. Math. Soc. 1961, 186–193 (MR 23 #A4104).

[9] Invariant distances on complex manifolds and holomorphic mappings. J. Math. Soc. Japan **19**, 460–480 (1967) (MR 38 #736).

[10] Hyperbolic manifolds and holomorphic mappings. New York: Marcel Dekker 1970.

Kobayashi, S., Nagano, T.
[1] On projective connections. J. Math. Mech. **13**, 215–236 (1964) (MR 28 #2501).
[2] On a fundamental theorem of Weyl-Cartan on *G*-structures. J. Math. Soc. Japan **17**, 84–101 (1965) (MR 33 #663).
[3] On filtered Lie algebras and geometric structures. I. J. Math. Mech. **13**, 875–908 (1964) (MR 29 #5961); II. **14**, 513–522 (1965) (MR 32 #2512); III. **14**, 679–706 (1965) (MR 32 #5803); IV. **15**, 163–175 (1966) (MR 33 #4189); V. **15**, 315–328 (1966) (MR 33 #4189).
[4] A report on filtered Lie algebras. Proc. US-Japan Seminar in Differential Geometry, Kyoto, 1965, 63–70 (MR 35 #6722).
[5] Riemannian manifolds with abundant isometries. Diff. Geometry in honor of K. Yano, Kinokuniya, Tokyo, 1972, 195–220.

Kobayashi, S., Nomizu, K.
[1] Foundations of differential geometry. New York: John Wiley & Sons, vol. 1, 1963 (MR 27 #2945); vol. 2, 1969 (MR 38 #6501).
[2] On automorphisms of a Kahlerian structure. Nagoya Math. J. **11**, 115–124 (1957) (MR 20 #4004).

Kobayashi, S., Ochiai, T.
[1] *G*-structures of order two and transgression operators. J. Differential Geometry **6**, 213–230 (1971).
[2] Mappings into compact complex manifolds with negative first Chern class. J. Math. Soc. Japan **23**, 137–148 (1971).

Kodaira, K.
[1] On differential geometric method in the theory of analytic stacks. Proc. Nat. Acad. Sci. U.S.A. **39**, 1263–1273 (1953) (MR 16, 618).
[2] On Kähler varieties of restricted type. Ann. of Math. **60**, 28–48 (1954) (MR 16, 252).
[3] On deformations of some complex pseudogroup structures. Ann. of Math. **71**, 224–302 (1960) (MR 22 #599).

Kosmann, Y.
[1] Dérvées de Lie des spineurs. C. R. Acad. Sci. Paris, Sér. A-B **262**, A 289–292 (1966) (MR 34 #723); Applications **262**, A 394–397 (1966) (MR 34 #724).

Kostant, B.
[1] Holonomy and the Lie algebra of infinitesimal motions of a Riemannian manifold. Trans. Amer. Math. Soc. **80**, 528–542 (1955) (MR 18, 930).
[2] On holonomy and homogeneous spaces. Nagoya Math. J. **12**, 31–54 (1957) (MR 21 # 6003).
[3] A characterization of invariant affine connections. Nagoya Math. J. **16**, 35–50 (1960) (MR 22 #1863).

Koszul, J. L.
[1] Sur la forme hermitienne canonique des espaces homogènes complexes. Canad. J. Math. **7**, 562–576 (1955) (MR 17, 1109).
[2] Domaines bornés homogènes et orbites de groupes de transformations affines. Bull. Soc. Math. France **89**, 515–533 (1961) (MR 26 #3090).
[3] Lectures on groups of transformations. Lectures on Math. & Physics, No. 20, Tata Inst. Bombay, 1965 (MR 36 #1571).

Ku, H. T., Mann, L. N., Sicks, J. L., Su, J. C.
[1] Degree of symmetry of a product manifold. Trans. Amer. Math. Soc. **146**, 133–149 (1969) (MR 40 #3579).

Kuiper, N.H.

[1] On conformally flat spaces in the large. Ann. of Math. **50**, 916–924 (1949) (MR 11, 133).

[2] On compact conformally Euclidean spaces of dimension >2. Ann. of Math. **52**, 478–490 (1950) (MR 12, 283).

[3] Groups of motions of order $\frac{1}{2}n(n-1)+1$ in Riemannian n-spaces. Indag. Math. **18**, 313–318 (1955) (MR 18, 232).

Kuiper, N.H., Yano, K.

[1] Two algebraic theorems with applications. Indag. Math. **18**, 319–328 (1956) (MR 18, 5).

Kulkarni, R.S.

[1] Curvature structures and conformal transformations. J. Differential Geometry **4**, 425–452 (1970).

[2] On a theorem of F. Schur, J. Differential Geometry **4**, 453–456 (1970).

Kuranishi, M.

[1] On the local theory of continuous infinite pseudogroups. I. Nagoya Math. J. **15**, 225–260 (1959) (MR 22 #6866); II. **19**, 55–91 (1961) (MR 26 #263).

Kurita, M.

[1] A note on conformal mappings of certain Riemannian manifolds. Nagoya Math. J. **21**, 111–114 (1962) (MR 26 #718).

[2] On the isometry of a homogeneous Riemann space. Tensor **3**, 91–100 (1954) (MR 16, 72).

Ledger, A.J., Obata, M.

[1] Compact riemannian manifolds with essential groups of conformorphisms. Trans. Amer. Math. Soc. **150**, 645–651 (1970) (MR 41 #7576).

Lee, E.H.

[1] On even-diemsnional skew symmetric spaces and their groups of transformations. Amer. J. Math. **67**, 321–328 (1945).

Lefebvre, J.

[1] Propriétés de groupe des transformations conformes et du groupe des automorphismes d'une variété localement conformement symplectique. C. R. Acad. Sci. Paris Sér. A-B **268**, A 717–719 (1969) (MR 39 #6199).

[2] Propriétés des algèbres d'automorphismes de certaines structures presque symplectiques. C. R. Acad. Sci. Paris Sér. A-B **266**, A 354–356 (1968) (MR 37 #6883).

[3] Transformations conformes et automorphismes de certaines structures presque symplectiques. C. R. Acad. Sci. Paris Sér. A-B **262**, A 752–754 (1966) (MR 34 #1954).

Lehmann-Lejeune, J.

[1] Intégrabilité des G-structures définies par une 1-forme 0-déformable à valeurs dans le fibre tangent. Ann. Inst. Fourier (Grenoble) **16**, 329–387 (1966) (MR 35 #3586).

Lehner, J., Newman, M.

[1] On Riemann surfaces with maximal automorphism groups. Glasgow Math. J. **8**, 102–112 (1967) (MR 36 #3976).

Lelong, P.

[1] Intégration sur un ensemble analytique complexe. Bull. Soc. Math. France **85**, 239–262 (1957) (MR 20 #2465).

Lelong-Ferrand, J.
[1] Quelques propriétés des groupes de transformations infinitésimales d'une variété riemannienne. Bull. Soc. Math. Belg. **8**, 15–30 (1956) (MR 19, 168).
[2] Sur les groupes à un parametre de transformations des variétés différentiables. J. Math. Pures Appl. **37**, 269–278 (1958) (MR 20 #4874).
[3] Sur l'application des méthodes de Hilbert à l'étude des transformations infinitésimales d'une variété différentiable. Bull. Soc. Math. Belg. **9**, 59–73 (1957) (MR 21 #2274).
[4] Transformations conformes et quasiconformes des variétés riemanniennes; applications à la démonstration d'une conjecture de A. Lichnerowicz. C. R. Acad. Sci. Paris Sér. A-B **269**, A 583–586 (1969) (MR 40 #7989).

Leslie, J. A.
[1] On a differential structure for the group of diffeomorphisms. Topology **6**, 263–271 (1967) (MR 35 #1041).
[2] Two classes of classical subgroups of Diff(M). J. Differential Geometry **5**, 427–435 (1971).

Lewittes, J.
[1] Automorphisms of compact Riemann surfaces. Amer. J. Math. **85**, 734–752 (1963) (MR 28 #4102).

Libermann, P.
[1] Sur le problème d'équivalence de certaines structures infinitésimales. Ann. Mat. Pura Appl. **36**, 27–120 (1954) (MR 16, 520).
[2] Sur les automorphismes infinitésimaux des structures symplectiques et des structures contactes, Colloque Géométrie Diff. Globale, Bruxelles, 1958, 37–59 (MR 22 #9919).
[3] Pseudogroupes infinitésimaux attachés aux pseudogroupes de Lie. Bull. Soc. Math. France **87**, 409–425 (1959) (MR 23 #A 607).

Lichnerowicz, A.
[1] Géométrie des groupes de transformations. Paris: Dunod 1958 (MR 23 #A 1329).
[2] Isométrie et transformations analytiques d'une variété kählérienne compacte. Bull. Soc. Math. France **87**, 427–437 (1959) (MR 22 #5012).
[3] Sur les transformations analytiques d'une variété kählérienne compacte. Colloque Geom. Diff. Global, Bruxelles, 1958, 11–26 (MR 22 #7150).
[4] Théorèmes de réductivité sur des algèbres d'automorphismes. Rend. Mat. e Appl. **22**, 197–244 (1963) (MR 28 #544).
[5] Variétés complexes et tenseur de Bergmann. Ann. Inst. Fourier (Grenoble) **15**, 345–407 (1965) (MR 33 #744).
[6] Variétés kählériennes et première classe de Chern. J. Differential Geometry **1**, 195–224 (1967) (MR 37 #2150).
[7] Transformations des variétés à connexion linéaire et des variétés riemannienne. Enseignement Math. **8**, 1–15 (1962) (MR 26 #717).
[8] Zéros des vecteurs holomorphes sur une variété kählérienne a première classe de Chern non-négative. Diff. Geometry in honor of K. Yano, Kinokuniya, Tokyo, 1972, 253–266.
[9] Variétés kählériennes à première classe de Chern non-négative et variétés riemanniennes à courbure de Ricci généralisée non-négative. J. Differential Geometry **6**, 47–94 (1971).

Loos, O.
[1] Symmetric spaces, 2 volumes. New York: Benjamin 1969 (MR 39 #365).

Lusztig, G.
[1] A property of certain non-degenerate holomorphic vector fields. An. Univ. Timişoara, Ser. Şti. Mat.-Fiz. **7**, 73–77 (1969) (MR 42 #2032).

Macbeath, A. M.
[1] On a theorem of Hurwitz. Proc. Glasgow Math. Assoc. **5**, 90–96 (1961) (MR 26 #4244).

Mann, L. N.
[1] Gaps in the dimensions of transformation groups. Illinois J. Math. **10**, 532–546 (1966) (MR 34 #282).

Manturov, O. V.
[1] Homogeneous Riemannian spaces with an irreducible rotation group. Trudy Sem. Vektor. Tenzor. Anal. **13**, 68–145 (1966) (MR 35 #926).

Matsumura, H.
[1] On algebraic groups of birational transformations. Accad. Naz. dei Lincei **34**, 151–155 (1963) (MR 28 #3041).

Matsumura, H., Monsky, P.
[1] On the automorphisms of hypersurfaces. J. Math. Kyoto Univ. **3-3**, 347–361 (1964) (MR 29 #5819).

Matsushima, Y.
[1] Sur la structure du groupe d'homéomorphismes analytiques d'une certaine variété kählérienne. Nagoya Math. J. **11**, 145–150 (1957) (MR 20 #995).
[2] Sur les espaces homogènes kählériens d'un groupe de Lie réductif. Nagoya Math. J. **11**, 53–60 (1957) (MR 19, 315).
[3] Sur certaines variétés homogènes complexes. Nagoya Math. J. **18**, 1–12 (1961) (MR 25 #2147).
[4] Holomorphic vector fields and the first Chern class of a Hodge manifold. J. Differential Geometry **3**, 477–480 (1969) (MR 42 #843).
[5] Holomorphic vector fields on compact Kähler manifolds. Conf. Board Math. Sci. Regional Conf. Ser. in Math. No. 7, 1971, Amer. Math. Soc.
[6] On Hodge manifolds with zero first Chern class. J. Differential Geometry **3**, 477–480 (1969).

Milnor, J.
[1] Lectures on Morse theory. Annals Math. Studies No. 51. Princeton Univ. Press 1963 (MR 29 #634).

Montgomery, D., Samelson, H.
[1] Transformation groups of spheres. Ann. of Math. **44**, 454–470 (1943) (MR 5, 60).

Montgomery, D., Zippin, L.
[1] Transformation groups. Interscience tracts #1. J. Wiley & Sons 1955 (MR 17, 383).

Morimoto, A.
[1] Sur le groupe d'automorphismes d'un espace fibré principal analytique complexe. Nagoya Math. J. **13**, 157–178 (1958) (MR 20 #2474).
[2] Prolongations of G-structures to tangent bundles. Nagoya Math. J. **32**, 67–108 (1968) (MR 37 #6865).
[3] Prolongations of G-structures to tangent bundles of higher order. Nagoya Math. J. **38**, 153–179 (1970) (MR 41 #7568).
[4] Prolongations of connections to tangential fibre bundles of higher order. Nagoya Math. J. **40**, 85–97 (1970).

Morimoto, A., Nagano, T.

[1] On pseudo-conformal transformations of hypersurfaces. J. Math. Soc. Japan **15**, 289–300 (1963) (MR 27 #5275).

Morimoto, T., Tanaka, N.

[1] The classification of the real primitive infinite Lie algebras. J. Math. Kyoto Univ. **10-2**, 207–243 (1970).

Moser, J.

[1] On the volume element on a manifold. Trans. Amer. Math. Soc. **120**, 286–294 (1965) (MR 32 #409).

Mutō, Y.

[1] On n-dimensional projectively flat spaces admitting a group of affine motions G_r of order $r > n^2 - n$. Sci. Rep. Yokohama Nat. Univ. Sec. I **4**, 3–18 (1955) (MR 17, 783).

[2] On the curvature affinor of an affinely connected manifold A_n, n 7 admitting a group of affine motions G_r of order $r > n^2 - 2n$. Tensor **5**, 39–53 (1955) (MR 17, 407).

Myers, S. B., Steenrod, N.

[1] The group of isometries of a Riemannian manifold. Ann. of Math. **40**, 400–416 (1939).

Nagano, T.

[1] The conformal transformation on a space with parallel Ricci tensor. J. Math. Soc. Japan **11**, 10–14 (1959) (MR 23 #A 1330).

[2] On conformal transformations in Riemannian spaces. J. Math. Soc. Japan **10**, 79–93 (1958) (MR 22 #1859).

[3] Sur des hypersurfaces et quelques groupes d'isometries d'un espace riemannien. Tôhoku Math. J. **10**, 242–252 (1958) (MR 21 #344).

[4] Transformation groups with $(n-1)$-dimensional orbits on noncompact manifolds. Nagoya Math. J. **14**, 25–38 (1959) (MR 21 #3513).

[5] On some compact transformation groups on spheres. Sci. Papers College Gen. Ed. Univ. Tokyo **9**, 213–218 (1959) (MR 22 #4799).

[6] Homogeneous sphere bundles and the isotropic Riemann manifolds. Nagoya Math. J. **15**, 29–55 (1959) (MR 21 #7522).

[7] Transformation groups on compact symmetric spaces. Trans. Amer. Math. Soc. **118**, 428–453 (1965) (MR 32 #419).

[8] The projective transformation on a space with parallel Ricci tensor. Kōdai Math. Sem. Rep. **11**, 131–138 (1959) (MR 22 #216).

[9] Isometries on complex-product spaces. Tensor **9**, 47–61 (1959) (MR 21 #6599).

[10] Linear differential systems with singularities and application to transitive Lie algebras. J. Math. Soc. Japan **18**, 398–404 (1966) (MR 33 #8005).

[11] 1-forms with the exterior derivatives of maximal rank. J. Differential Geometry **2**, 253–264 (1968) (MR 39 #883).

Nakano, S.

[1] On complex analytic vector bundles. J. Math. Soc. Japan **7**, 1–12 (1955) (MR 17, 409).

Narasimhan, M. S., Simha, R. R.

[1] Manifolds with ample canonical class. Invent. Math. **5**, 120–128 (1968) (MR 38 #5253).

Naruki, I.

[1] A note on the automorphism group of an almost complex structure of type (n, n'). Proc. Japan Acad. **44**, 243–245 (1968) (MR 37 #3586).

Newlander, A., Nirenberg, L.

[1] Complex analytic coordinates in almost complex manifolds. Ann. of Math. **65**, 391–404 (1957) (MR 19, 577).

Nijenhuis, A., Woolf, W. B.

[1] Some integration problems in almost complex manifolds. Ann. of Math. **77**, 424–483 (1963) (MR 26 #6992).

Nomizu, K.

[1] On the group of affine transformations of an affinely connected manifold. Proc. Amer. Math. Soc. **4**, 816–823 (1953) (MR 15, 468).

[2] Invariant affine connections on homogeneous spaces. Amer. J. Math. **76**, 33–65 (1954) (MR 15, 468).

[3] Lie groups and differential geometry. Publ. Math. Soc. Japan #2, 1956 (MR 18, 821).

[4] On local and global existence of Killing vector fields. Ann. of Math. **72**, 105–120 (1960) (MR 22 #9938).

[5] Sur les algèbres de Lie de générateurs de Killing et l'homogénéité d'une variété riemannienne. Osaka Math. J. **14**, 45–51 (1962) (MR 25 #4467).

[6] Holonomy, Ricci tensor and Killing vector fields. Proc. Amer. Math. Soc. **12**, 594–597 (1961) (MR 24 #A 1099).

[7] Recent development in the theory of connections and holonomy groups. Advances in Math., fasc. 1, 1–49 New York: Academic Press (MR 25 #5473).

Nomizu, K., Yano, K.

[1] On infinitesimal transformations preserving the curvature tensor field and its covariant differentials. Ann. Inst. Fourier (Grenoble) **14**, 227–236 (1964) (MR 30 #4227).

[2] Some results related to the equivalence problem in Riemannian geometry. Proc. US-Japan Seminar in Diff. Geometry, Kyoto, Japan, 1965, 95–100 (MR 37 #2136); Math. Z. **97**, 29–37 (1967) (MR 37 #2137).

Obata, M.

[1] Affine transformations in an almost complex manifold with a natural affine connection. J. Math. Soc. Japan **8**, 345–362 (1956) (MR 18, 822).

[2] On the subgroups of the orthogonal groups. Trans. Amer. Math. Soc. **87**, 347–358 (1958) (MR 20 #1711).

[3] Certain conditions for a Riemannian manifold to be isometric with a sphere. J. Math. Soc. Japan **14**, 333–340 (1962) (MR 25 #5479).

[4] Conformal transformations of compact Riemannian manifolds. Illinois J. Math. **6**, 292–295 (1962) (MR 25 #1507).

[5] Riemannian manifolds admitting a solution of a certain system of differential equations. Proc. US-Japan Seminar in Diff. Geometry, Kyoto, Japan, 1965, 101–114 (MR 35 #7263).

[6] Conformal transformations of Riemannian manifolds. J. Differential Geometry **4**, 331–334 (1970) (MR 42 #2387).

[7] Conformally flat Riemannian manifolds admitting a one-parameter group of conformal transformations. J. Differential Geometry **4**, 335–338 (1970) (MR 42 #2388).

[8] The conjectures on conformal transformations of Riemannian manifolds. Bull. Amer. Math. Soc. **77**, 265–270 (1971) (MR 42 #5286). J. Differential Geometry **6**, 247–258 (1971).

[9] Conformal changes of Riemannian metrics on a Euclidean sphere. Diff. Geometry in honor of K. Yano, Kinokuniya, Tokyo, 345–354 (1972).

[10] On n-dimensional homogeneous spaces of Lie groups of dimension greater than $n(n-1)/2$. J. Math. Soc. Japan **7**, 371–388 (1955) (MR 18, 599).

Obata, M., Ledger, A. J.
[1] Transformations conformes des variétés riemanniennes compactes. C. R. Acad. Sci. Paris Sér. A-B **270**, A459–461 (1970) (MR 40 #7990).

Ochiai, T.
[1] Classification of the finite nonlinear primitive Lie algebras. Trans. Amer. Math. Soc. **124**, 313–322 (1966) (MR 34 #4320).
[2] On the automorphism group of a *G*-structure. J. Math. Soc. Japan **18**, 189–193 (1966) (MR 33 #3224).
[3] Geometry associated with semi-simple flat homogeneous spaces. Trans. Amer. Math. Soc. **152**, 1–33 (1970).

Oeljeklaus, E.
[1] Über fasthomogene kompakte Mannigfaltigkeiten. Schr. Math. Inst. Univ. Münster (2) Heft 1 (1970) (MR 42 #2033).

Ogiue, K.
[1] Theory of conformal connections. Kōdai Math. Sem. Rep. **19**, 193–224 (1967) (MR 36 #812).
[2] *G*-structures of higher order. Kōdai Math. Sem. Rep. **19**, 488–497 (1967) (MR 36 #814).

Omori, H.
[1] On the group of diffeomorphisms on a compact manifold. Proc. Symp. Pure Math. XV, Amer. Math. Soc., 1970, 167–183 (MR 42 #6864).
[2] Groups of diffeomorphisms and their subgroups, to appear.
[3] On contact transformation groups. J. Math. Soc. Japan, to appear.

Ozols, V.
[1] Critical points of the displacement functions of an isometry. J. Differential Geometry **3**, 411–432 (1969) (MR 42 #1010).

Palais, R.
[1] A global formulation of the Lie theory of transformation groups. Mem. Amer. Math. Soc. #22, 1957 (MR 22 #12162).
[2] On the differentiability of isometries. Proc. Amer. Math. Soc. **8**, 805–807 (1957) (MR 19, 451).

Peters, K.
[1] Über holomorphe und meromorphe Abbildungen gewisser kompakter komplexer Mannigfaltigkeiten. Arch. Math. (Basel) **15**, 222–231 (1964) (MR 29 #4071).

Pjatetzki-Shapiro, I. I.
[1] Géométrie des domaines classiques et théorie des fonctions automorphes. Paris: Dunod 1966 (MR 33 #5949).

Poincaré, H.
[1] Sur un théorème de M. Fuchs. Acta Math. **7**, 1–32 (1885).

Preismann, A.
[1] Quelques propriétés globales des espaces de Riemann. Comment. Math. Helv. **15**, 175–216 (1942) (MR 6, 20).

Raševskiĭ, P. K.
[1] On the geometry of homogeneous spaces. Dokl. Akad. Nauk SSSR **80**, 169–171 (1951) (MR 13, 383); Trudy Sem. Vektor. Tenzor. Anal. **9**, 49–74 (1952) (MR 14, 795).

Remmert, R.
[1] Holomorphe und meromorphe Abbildungen komplexer Räume. Math. Ann. **133**, 328–370 (1957) (MR 19, 1193).

Rodrigues, A. M.
[1] The first and second fundamental theorems of Lie for Lie pseudo groups. Amer. J. Math. **84**, 265–282 (1962) (MR 26 #5102).
[2] On Cartan pseudo groups. Nagoya Math. J. **23**, 1–4 (1963) (MR 30 #4862).

Rossi, H.
[1] Vector fields on analytic spaces. Ann. of Math. **78**, 455–467 (1963) (MR 29 #277).

Ruh, E. A.
[1] On the automorphism groups of a G-structure. Comment. Math. Helv. **39**, 189–264 (1964) (MR 31 #1631).

Samelson, H.
[1] Topology of Lie groups. Bull. Amer. Math. Soc. **58**, 2–37 (1952) (MR 13 #533).

Sampson, H.
[1] A note on automorphic varieties. Proc. Nat. Acad. Sci. U.S.A. **38**, 895–898 (1952) (MR 14, 633).

Sasaki, S.
[1] Almost contact manifolds. Math. Inst. Tôhoku Univ., Part I, Part II, 1967; Part III, 1968.

Sawaki, S.
[1] A generalization of Matsushima's theorem. Math. Ann. **146**, 279–286 (1962) (MR 25 #538).

Schwarz, H. A.
[1] Über diejenigen algebraischen Gleichungen zwischen zwei veränderlichen Größen, welche eine Schar rationaler eindeutig umkehrbarer Transformationen in sich selbst zulassen. Crelle's J. **87**, 139–145 (1879).

Shnider, S.
[1] The classification of real primitive infinite Lie algebras. J. Differential Geometry **4**, 81–89 (1970).

Singer, I. M.
[1] Infinitesimally homogeneous spaces. Comm. Pure Appl. Math. **13**, 685–697 (1960) (MR 24 #A 1100).

Singer, I. M., Sternberg, S.
[1] On the infinite groups of Lie and Cartan. I. Ann. Inst. Fourier (Grenoble) **15**, 1–114 (1965) (MR 36 #991).

Ślebodziński, A.
[1] Formes extérieurs et leurs applications. Warzawa, vol. 1, 1954 (MR 16, 1082); vol. 2, 1963 (MR 29 #2738).

Solodovnikov,
[1] The group of projective transformations in a complete analytic Riemannian space. Dokl. Akad. Nauk SSSR **186**, 1262–1265 (1969); Soviet Math. Dokl. **10**, 750–753 (1969) (MR 40 #1936).

Spencer, D. C.
[1] Deformation of structures on manifolds defined by transitive, continuous pseudogroups. Part I. Ann. of Math. **76**, 306–398 (1962); Part II, 399–445 (MR 27 #6287).

Sternberg, S.
[1] Lectures on differential geometry. Prentice-Hall 1964 (MR 23 #1797).

Suyama, Y., Tsukamoto, Y.
[1] Riemannian manifolds admitting a certain conformal transformation group. J. Differential Geometry **5**, 415–426 (1971).

Synge, J. L.
[1] On the connectivity of spaces of positive curvature. Quart. J. Math. **7**, 316–320 (1936).

Takeuchi, M.
[1] On the fundamental group and the group of isometries of a symmetric space. J. Fac. Sci. Univ. Tokyo Sect. I **10**, 88–123 (1964) (MR 30 #1217).

Takizawa, S.
[1] On contact structures of real and complex manifolds. Tôhoku Math. J. **15**, 227–252 (1963) (MR 28 #566).

Tanaka, N.
[1] Conformal connections and conformal transformations. Trans. Amer. Math. Soc. **92**, 168–190 (1959) (MR 23 A 1331).
[2] Projective connections and projective transformations. Nagoya Math. J. **11**, 1–24 (1957) (MR 21 #3899).
[3] On the pseudo-conformal geometry of hypersurfaces of the space of *n* complex variables. J. Math. Soc. Japan **14**, 397–429 (1962) (MR 26 #3086).
[4] On the equivalence problems associated with a certain class of homogeneous spaces. J. Math. Soc. Japan **17**, 103–139 (1965) (MR 32 #6358).
[5] Graded Lie algebras and geometric structures. Proc. US-Japan Seminar in Diff. Geometry, Kyoto, 1965, 147–150 (MR 36 #5852).
[6] On generalized graded Lie algebras and geometric structures. I. J. Math. Soc. Japan **19**, 215–254 (1967) (MR 36 #4470).
[7] On differential systems, graded Lie algebras and pseudo-groups. J. Math. Kyoto Univ. **10**, 1–82 (1970) (MR 42 #1165).
[8] On infinitesimal automorphisms of Siegel domains. J. Math. Soc. Japan **22**, 180–212 (1970) (MR 42 #7939).

Tanno, S.
[1] Strongly curvature-preserving transformations of pseudo-Riemannian manifolds. Tôhoku Math. J. **19**, 245–250 (1967) (MR 36 #3284).
[2] Transformations of pseudo-Riemannian manifolds. J. Math. Soc. Japan **21**, 270–281 (1969) (MR 39 #2101).
[3] The automorphism groups of almost Hermitian manifolds. Trans. Amer. Math. Soc. **137**, 269–275 (1969) (MR 38 #5144).

Tanno, S., Weber, W. C.
[1] Closed conformal vector fields. J. Differential Geometry **3**, 361–366 (1969) (MR 41 #6111).

Tashiro, Y.
[1] Complete Riemannian manifolds and some vector fields. Trans. Amer. Math. Soc. **117**, 251–271 (1965) (MR 30 #4229).
[2] Conformal transformations in complete Riemannian manifolds. Publ. Study Group of Geometry, vol. 3, Kyoto Univ. 1967 (MR 38 #639).
[3] On projective transformations of Riemannian manifolds. J. Math. Soc. Japan **11**, 196–204 (1959) (MR 23 #A 602).

Tashiro, Y., Miyashita, K.
[1] Conformal transformations in complete product Riemannian manifolds. J. Math. Soc. Japan **19**, 328–346 (1967) (MR 35 #3605).
[2] On conformal diffeomorphisms of complete Riemannian manifolds with parallel Ricci tensor. J. Math. Soc. Japan **23**, 1–10 (1971).

Teleman, C.
[1] Sur les groupes maximums de mouvement des espaces de Riemann V_n. Acad. R. P. Romîne Stud. Cerc. Mat. **5**, 143–171 (1954) (MR 16, 624).
[2] Sur les groupes des mouvements d'un espace de Riemann. J. Math. Soc. Japan **15**, 134–158 (1963) (MR 27 #6219).

Veisfeiler, B. Ju.
[1] On filtered Lie algebras and their associated graded algebras. Funkcional. Anal. i Priložen **2**, 94 (1968).

Vranceanu, G.
[1] Leçons de Géométrie Différentielle, 3 volumes. Gauthier-Villars 1964.

Wakakuwa, H.
[1] On n-dimensional Riemannian spaces admitting some groups of motions of order less than $n(n-1)/2$. Tôhoku Math. J. **6**, 121–134 (1954) (MR 16, 956).

Wang, H.C.
[1] Finsler spaces with completely integrable equations of Killing. J. London Math. Soc. **22**, 5–9 (1947) (MR 9, 206).
[2] On invariant connections over a principal fibre bundle. Nagoya Math. J. **13**, 1–19 (1958) (MR 21 #6001).
[3] Two point homogeneous spaces. Ann. of Math. **55**, 177–191 (1952) (MR 13, 863).
[4] Closed manifolds with homogeneous complex structure. Amer. J. Math. **76**, 1–32 (1954) (MR 16, 518).

Wang, H.C., Yano, K.
[1] A class of affinely connected spaces. Trans. Amer. Math. Soc. **80**, 72–96 (1955) (MR 17, 407).

Weber, W.C., Goldberg, S.I.
[1] Conformal deformations of riemannian manifolds. Queen's paper in pure & applied Math., No. 16, Queen's Univ. Kingston, Ont. 1969 (MR 40 #1938).

Weil, A.
[1] Introduction à l'Étude des Variétés Kählériennes. Paris: Hermann 1958 (MR 22 #1921).

Weinstein, A.
[1] A fixed point theorem for positively curved manifolds. J. Math. Mech. **18**, 149–153 (1968) (MR 37 #3478).
[2] Symplectic structures on Banach manifolds. Bull. Amer. Math. Soc. **75**, 1040–1041 (1969).

Weyl, H.
[1] Reine Infinitesimalgeometrie. Math. Z. **2**, 384–411 (1918).
[2] Zur Infinitesimalgeometrie; Einordnung der projektiven und konformen Auffassung. Göttinger Nachrichten (1912), 99–112.

Wolf, J.A.
[1] Spaces of constant curvature. New York: McGraw-Hill 1967 (MR 36 #829).
[2] Homogeneity and bounded isometries in manifolds of negative curvature. Illinois J. Math. **8**, 14–18 (1964) (MR 29 #565).

[3] The geometry and structures of isotropy irreducible homogeneous spaces. Acta Math. **120**, 59–148 (1968) (MR 36 #6549).
[4] Growth of finitely generated solvable groups and curvature of riemannian manifolds. J. Differential Geometry **2**, 421–446 (1968) (MR 40 #1939).
[5] The automorphism group of a homogeneous almost complex manifold. Trans. Amer. Math. Soc. **144**, 535–543 (1969) (MR 41 #956).

Wright, E.
[1] Thesis, Univ. of Notre Dame.

Wu, H.
[1] Normal families of holomorphic mappings. Acta Math. **119**, 193–233 (1967) (MR 37 #468).

Yano, K.
[1] On harmonic and Killing vector fields. Ann. of Math. **55**, 38–45 (1952) (MR 13, 689).
[2] On n-dimensional Riemannian spaces admitting a group of motions of order $\frac{1}{2}n(n-1)+1$. Trans. Amer. Math. Soc. **74**, 260–279 (1953) (MR 14, 688).
[3] The theory of Lie derivatives and its applications. Amsterdam: North-Holland, 1957 (MR 19, 576).
[4] Sur un théorème de M. Matsushima. Nagoya Math. J. **12**, 147–150 (1957) (MR 20 #2476).
[5] Harmonic and Killing vector fields in compact orientable manifolds Riemannian spaces with boundary. Ann. of Math. **69**, 588–597 (1959) (MR 21 #3887).
[6] Some integral formulas and their applications. Michigan Math. J. **5**, 68–73 (1958) (MR 21 #1860).
[7] Differential geometry on complex and almost complex spaces. New York: Pergamon Press 1965 (MR 32 #4635).
[8] On Riemannian manifolds admitting an infinitesimal conformal transformations. Math. Z. **113**, 205–214 (1970) (MR 41 #6114).
[9] Integral formulas in Riemannian geometry. New York: Marcel Dekker 1970.

Yano, K., Bochner, S.
[1] Curvature and Betti numbers. Ann. of Math. Studies, No. 32. Princeton Univ. Press 1953 (MR 15, 989).

Yano, K., Kobayashi, S.
[1] Prolongations of tensor fields and connections to tangent bundles. J. Math. Soc. Japan **18**, 194–210 (1966) (MR 33 #1816); II. 236–246 (MR 34 #743); III. **19**, 486–488 (1967) (MR 36 #2084).

Yano, K., Nagano, T.
[1] Einstein spaces admitting a one-parameter group of conformal transformations. Ann. of Math. **69**, 451–461 (1959) (MR 21 #345).
[2] Some theorems on projective and conformal transformations. Indag. Math. **19**, 451–458 (1957) (MR 22 #1861).
[3] The de Rham decomposition, isometries and affine transformations in Riemannian spaces. Japan. J. Math. **29**, 173–184 (1959) (MR 22 #11341).

Yano, K., Obata, M.
[1] Conformal changes of Riemannian metrics. J. Differential Geometry **4**, 53–72 (1970) (MR 41 #6113).

Yau, S.T.
[1] On the fundamental group of compact manifolds of non-positive curvature. Ann. of Math. **93**, 579–585 (1971) (MR 42 #8423).

Index